Thank you!

I appreciate your support of Adamantine Energy and my work.

It's my pleasure to present you with this copy of my new book!

As I launch this work into the world, your voice and support mean everything.

Ways you can contribute to the success of the book launch:

1. **Leave a short, honest book review on Amazon.** Every review helps.
2. **Share the book with your network**—on LinkedIn, on Substack, or in conversation. (I've included sample messages below to make it easy.)
3. **Buy additional copies** for your team or board: https://energythinks.com/the-myth-and-the-moment/. For bulk pricing, contact lindsey@energythinks.com.
4. **Recommend the book to three industry peers** who are ready to lead in this moment.

Sample social media posts:

- Excellent new book: *The Myth and The Moment* by @Tisha Schuller. A must-read for industry leaders. https://energythinks.com/the-myth-and-the-moment/
- I love this new book by my colleague @Tisha Schuller; get a copy of *The Myth and The Moment* here: https://energythinks.com/the-myth-and-the-moment/
- A great read for anyone navigating today's energy landscape. Thanks, @Tisha Schuller, for *The Myth and The Moment*! https://energythinks.com/the-myth-and-the-moment/
- Just finished *The Myth and The Moment* by @Tisha Schuller—smart, pragmatic, and perfectly timed. I highly recommend it! https://energythinks.com/the-myth-and-the-moment/

Thank you again for all the ways you continue to support me and Adamantine. I hope this book serves as a valuable tool for your leadership journey.

With gratitude,
Tisha

"Tisha continues to be a refreshing voice in today's complex energy environment. *The Myth and The Moment* perfectly captures the opportunity we have as an industry to lead and bring others along with us."

—Alan Armstrong, *Executive Chairman, Williams Board of Directors*

"*The Myth and The Moment* breaks through the noise with a clear, compelling argument: The U.S. energy industry has a window of opportunity to shape lasting opinions and get something done. With characteristic candor and humor, Tisha provides a practical playbook for how industry can meet that moment."

—Amy Cradic, *COO and Senior Vice President, New Jersey Resources*

"Tisha delivers a practical and refreshingly honest playbook for oil and gas executives navigating an uncertain energy future."

—John Dabbar, *CSC Chair, National Petroleum Council Committee on GHG Emissions*

"Tisha's new book is a welcome addition to the discourse. It's sharp, provocative, and honest about the trade-offs for energy companies navigating the uncertain politics of our energy future."

—Arnab Datta, *Managing Director of Policy Implementation, Employ America*

"Tisha Schuller once again delivers a clear-eyed perspective and ready-for-action insights to help leaders find our next footing as we navigate through uncertainty."

—Paula Gant, *President & CEO, GTI Energy*

"*The Myth and The Moment* reflects the real-world clarity that our sector—and our communities—need right now. It's a coming-of-age guide for leaders ready to meet complexity with common sense."

—Scott Hallam, *President and CEO, Boardwalk Pipelines*

"*The Myth and The Moment* offers a timely and practical road map for executives navigating the dual challenge of delivering more energy to a growing world while reducing carbon emissions. In the face of an increasingly complex and shifting political landscape, Tisha Schuller distills clarity from chaos. Her approach is both strategic and accessible—empowering leaders to act with confidence and purpose."

—Mike Howard, *Chairman and CEO, Howard Energy Partners*

"Another brilliant, thought-provoking, and strategic guide for our oil and gas industry executives. It concisely lays out how we may reclaim our energy relevance through partnership, collaboration, and pragmatism. The choices are clear: either complacency with an "I told you so" attitude or true leadership that confirms the coexistence of energy abundance and decarbonization."

—Al Huehn, *President and CEO, Topco Oilsite Products Ltd.*

"In *The Myth and the Moment,* Tisha Schuller masterfully unpacks what many of us have felt coming: The green euphoria of the pandemic has faded, and investors, governments, and consumers are waking up with a five-year hangover. As Tisha puts it, "The Myth of an Easy Transition" has unraveled—and in its place lies a rare opportunity. Drawing from private conversations with global energy and climate leaders, she offers a pragmatic call to action: Oil and gas is back but will only lead the future of energy if we seize this moment. For those serious about delivering abundant, affordable, and sustainable energy—not as a slogan, but as a strategy—this book is an essential guide to leading with humility, courage, and conviction in the messy middle where real solutions live."

—Kevin Krausert, *CEO and Co-Founder, Avatar Innovations*

"At a time when our politics are as fraught as ever, The Myth and the Moment provides timely, pragmatic guidance for navigating today's energy and climate challenges toward a positive future of security and abundance."

—Roger Pielke Jr., *Senior Fellow, American Enterprise Institute*

"Tisha Schuller has created a pragmatic and inspirational path forward for all who care about the natural world and the prosperity of humankind. The collapse of the climate myth offers a fleeting moment for business, political, and community leaders to create a new reality around energy and climate."

—George Sparks, *President and CEO, Denver Museum of Nature & Science*

"Tisha delivers a practical and refreshingly honest playbook for oil and gas executives navigating an uncertain energy future."

—Dena Wiggins, *CEO and President of a U.S. natural gas trade association*

THE MYTH AND THE MOMENT

From Polarization to Progress in the New Energy Landscape

TISHA SCHULLER

Lodgepole Pine
PUBLISHING

The Myth and The Moment: From Polarization to Progress in the New Energy Landscape

ISBN: 978-0-9993220-3-1

To the unwritten future—
may our collaboration shape it for the better.

To those willing to stand up to their tribe—
that requires the real courage and heart.

TABLE OF CONTENTS

Preface

UNDERSTANDING, FOR ME, can dawn slowly. Before grasping an important new idea about the energy–climate dynamic, I feel an achy sense of disorientation: Something is not quite right; there is information I am missing; my once comfortable understanding of the balance and interactions between the ever-rising demands for energy and the public mandate for climate action feels itchy.

That's the space I was in for six months in 2024 as I researched and wrote about what I would come to call The Myth and The Moment. I was feeling out the puzzle pieces, but I wasn't quite ready to slide them into place. And then my husband and I met some family friends for dinner.

I didn't want to talk shop, but good Colorado liberals always want to press me on what I do for a living. They do it under the guise of "educating themselves," but it's always a thinly veiled indictment of my work with the oil and gas industry. "Tell me about your book" quickly becomes "It feels like we are in the apocalypse! We don't have any time to waste!" In this case, we were sharing tapas while wildfires in California were burning out of control, and so the transition to grilling me was quicker than usual.

Those of you who are familiar with my work may think I have infinite patience for and interest in explaining the complexities of the energy transition to well-meaning liberals. I don't. And even if I did, I was halfway through my second beer, which I had to push out of reach to ensure that I kept my wits about me. It was a Friday night, and this was the last thing I wanted to discuss. But I'm nothing if not dedicated to my cause … even as my fingers twitched closer to my glass.

The conversation covered the two big energy-and-climate conversational traps, at least for well-meaning liberals: (1) We are doomed, and we have to do everything within our power as quickly as possible to reduce climate emissions; and (2) it's companies that need to make the energy transition moves, and they need to be forced to do so by regulation (yesterday!). I did my best to explain that while these views are rooted in idealism, they are unworkable in practice.

The next morning, I had an emotional hangover from the conversation. I'd made a little impact but not much. My thoughts were too complicated, too nuanced. But I suddenly understood on a gut level just how important these two traps are, because they provide the backdrop for not just The Myth but The Moment.

Just a couple of weeks after that dinner, while I was still processing what had happened and trying to articulate The Myth—The Myth of an Easy Energy Transition—it basically landed in my lap.

I found myself on a two-person ski lift on the bunny hill as part of my quest to switch from snowboarding to skiing. (This was easier to tackle than The Myth, by the way.) The lift was painfully slow, and my fellow ski student was interested in my work. I'd barely gotten out of my mouth a description of what my firm, Adamantine Energy, does for oil and gas companies, "We build strategies on all things the future of energy—"

"Like what technologies, exactly?" she asked, in a way that made me wish I had caught a faster lift. "I'm a professor at CU Boulder, so I understand this space."

"Well, oil and gas–adjacent technologies like geothermal—" I began, but that's all I got out.

"Like CCS? DAC? You know I went to COP in Baku, Azerbaijan," she interrupted, "and those just aren't going to work. I know it's bad for your *clients*, for your *business*, but what we need is to get off all oil and gas *today*."

Why aren't we at the top yet?

Providing what turned out to be a bit of internal comic relief, she had just told me that she was worried because she had driven her brand-new electric vehicle up to the resort. It was both stuck in snow at the charger *and* too cold to charge. She had to sort out a tow and a ride. Now I thought, *Mm-hmm. Maybe that's why we might need oil and gas for a wee bit longer, to get you a tow?* But instead of saying that, I just looked at her, smiled, and stopped talking. *What could I possibly say that would help?* And perhaps stewing in her own bad manners would be good for her skiing.

I kid you not: The rest of the lesson she chased me around to continue the conversation. Fortunately, I'm a very fast learner, and she was unable to catch me on skis.

This experience added to my consternation about The Myth and The Moment. The conversation happened the week President Donald Trump was inaugurated for the second time, and I thought, on a subsequent (mercifully quiet) lift ride, *Doesn't she understand that it is this type of "liberal elite" righteous approach that turned off so many voters? That made so many oil and gas workers feel alienated and angry?*

No, she doesn't. And she might never. Because that's the power of The Myth.

So what about The Moment?

The Moment is the window of opportunity for energy leadership that opens when The Myth unravels.

If you're an energy leader, you already know all too well about The Myth—and odds are you already sense The Moment.

How will true climate believers—including many of your stakeholders—process The Myth's unraveling?

Early returns tell us it will be messy.

Here's what's clear: The Moment is emerging, fast—and you, the energy leader, will very soon need to choose whether to seize the opportunities it will afford you.

But to understand the opportunities of The Moment, you first have to understand how entrenched energy and climate tropes—and with them, The Myth—are in the minds and hearts of the true believers like my ski-lesson buddy. These individuals have built careers, identities, and entire organizations around a climate-centric worldview. And I'm not just talking about self-assured college professors. Many of The Myth's adherents are our customers, community members, investors, regulators, and even employees.

In other words: many of our most important stakeholders.

What seems obvious to us—the unravelling of The Myth—was not even on the radar of either my dinner companions or my ski-class mate. We have to understand what it will take for the true believers (who are also our stakeholders!) to reset and revise their worldview. Because it isn't just a shift in strategy for them; it's a seismic shift. A loss of purpose. Of certainty. Of a clear sense of right and wrong. Of good guys and bad guys. That's why they have held and will hold on so tightly to The Myth—and to the narrow set of solutions they've invested in for years.

Without our help, that is.

In early 2025, two things became undeniable: First, The Moment was arriving faster than I expected—so fast, in fact, that it was already here. Second, the most passionate proponents of The Myth weren't just unprepared for The Moment—they couldn't even see it. I kept thinking about that dinner, that ski lift, all those conversations with well-intentioned, smart, left-leaning folks. They spoke with urgency, with conviction, with moral certainty. And yet they were missing the cracks forming right beneath their feet.

The Moment exists because energy and climate have been swept up in our hyper-polarized political landscape. You (and I) see the world through an energy-centric lens. You (and I) think in terms of reliability, demand, innovation, and scale. But many of your stakeholders don't. Depending on where you work, somewhere between 20 percent and 60 percent of them hold a climate-centric worldview. For them, energy isn't the center of the story—climate is. They've all fallen into the two traps my companions fell into: (1) We're doomed on climate unless we act quickly, now; and (2) companies have to be forced through regulation to act quickly, now, on climate. And they fervently believe that the energy transition will be easy and painless (except, of course, for the oil and gas companies).

Everything I'm going to share in this book about The Myth and The Moment threatens that worldview. The energy transition is hard. It's messy. It's complicated. Consumers are going to have to both pay and participate. Incremental progress matters. Imperfect solutions will be necessary. Climate change will require both mitigation and adaptation. If governments overreact, they'll drive up costs or trigger blackouts—and the public (consumers!) will revolt, setting back progress. Regulation needs to be thoughtful and balanced. We're going to need *all* the solutions. There's no bad guy. There's no good guy.

After that ski lesson, I couldn't stop thinking about what this means for us—for oil and gas leaders. The Moment isn't theoretical. It's here. And it brings unprecedented opportunity to those in our industry who step up: opportunity to accelerate the work we're already doing to reduce carbon intensity, to expand our low-carbon business lines, and to lead in a space that's being completely redefined in real time.

Another reason to step up: Because while The Myth is unraveling, it's far from defeated. It's too deeply rooted in culture and belief systems to fade on its own. Without a viable alternative, it will snap right back into place when political tides turn. In that case, the progressive left—the Myth's biggest champions—will be back in charge. They'll return stronger, with more urgency and more blame—and oil and gas will be even further out in the cold.

My dinner and ski lift trip taught me this: We need a very particular kind of leadership as The Myth unravels, leadership that is in our DNA as an industry. Leading in The Moment doesn't mean retreating or fighting back; it means stepping into the middle, where real solutions are forged. It requires partnering with stakeholders who recognize that The Myth no longer works but are uncertain about what could possibly come next.

This is how we lead and solve the problem of the energy future.

Our solutions—on emissions, innovation, scale, decarbonization—come next. *We* come next.

But only if we engage properly.

The energy future depends entirely on how we engage. Not just internally, with our teams and strategies, but externally, with the people who have the most to lose from The Myth's unraveling. We're going to need new kinds of partnerships. Unlikely allies. And a healthy dose of humility.

That's why I wrote this book: Because what comes after The Moment isn't a foregone conclusion. The story hasn't been written. The outcomes are far from certain. You can imagine a world where four years from now, we'll have made no progress. Where you'll have built the tech, the strategy, even the narrative—but you'll still be the villain. Still stuck in the same tired debates.

But you're here, reading this, because you see things differently. You're a prescient leader, and you know the ground is shifting. This book is for you. And leaders like you—those who act now—will shape what comes next for oil and gas.

Buckle up!

PART I

What Is The Moment?

CHAPTER 1

The Myth

I believe that imagination is stronger than knowledge. That myth is more potent than history. That dreams are more powerful than facts. That hope always triumphs over experience.

—Robert Fulghum, *All I Really Need to Know I Learned in Kindergarten*

THE MOMENT EXISTS only because The Myth came first—The Myth of an Easy Energy Transition. Put forward in the 2010s by well-intentioned optimists, the Myth sold half the country (and much of the world) on the idea that emissions could and would drop quickly, driven by policy and the resulting technological breakthroughs. And this would lead to a relatively painless and eminently affordable transition to zero-greenhouse-gas emissions!

The Myth held sway for over a decade. But in The Moment, The Myth is unraveling. What happens when The Myth is completely undone?

In the fall of 2022, The Myth was at its peak. As an observer, I attended a focus group session in Wyoming—one of the most politically conservative states in the country—convened to gather input from community members on a hypothetical direct air capture (DAC) facility with carbon capture and sequestration (CCS).

Early in my career, I did a lot of fieldwork in Wyoming, and I love it there. I was an interested observer of this focus group, held by a pro-DAC team. The focus group was being held to assess whether a region that for over a century had produced coal, oil, and natural gas would be receptive to energy transition solutions such as DAC. (Spoiler alert: It depends.)

The team of five hosting the focus group had arrived a day early in Wyoming with a strong, often unconscious, liberal progressive bias. Over dinner, I learned that the team members assumed that the community would want the jobs and economic development. They spoke as if *all* communities wanted to transition away from fossil fuels, especially ones that had been subject to boom-and-bust cycles for decades. Although our faces were lit by the "Welcome to Coal Country" sign outside of the bar window, they weren't particularly interested in my cautious notes about their assumptions.

At the time, the spotlight for organizations like the one holding the focus group was on environmental justice (EJ). They had been working on improving outreach in urban, historically disadvantaged communities. To their credit, they understood that many decarbonization projects were being proposed in red communities and states, and this group—smartly—decided to test their EJ tool kit in a very conservative community.

What the team didn't understand was how the "justice" paradigm didn't really apply in Wyoming. Although many on the left have generous-spirited ideas about a "just transition" for oil and gas workers and their communities, I have yet to meet a worker or community that is interested in being "justly transitioned." The truest form of "just transition" will *include*—not sideline—oil and gas companies, workers, and communities in building the energy future. And engagement with these communities will not look like project engagement with urban or blue-leaning communities.

The next morning, 50 people gathered around tables, quietly waiting for the focus group to begin. I squirmed in the back of the room. The team presenting the hypothetical DAC project didn't understand the community they were speaking to—and it showed. Their assumptions were even more problematic than the ones I'd heard over dinner the night before.

For starters, the presenters didn't bother to explain *why* decarbonization was important—nor even necessary. They casually framed labor unions and community organizers as assets for project development. And they made it clear that traditional oil, gas, and mining companies wouldn't be involved—because, in their words, those industries had "done enough damage." The cues in the room were obvious, if you were paying attention: crossed arms, darting glances, the occasional harumph. But the presenters missed it all. When you're steeped in The Myth, it's hard to imagine a world where it doesn't dominate.

The reaction from the community was swift—and cutting. One person said, "You can take your fake-news, windmill-loving projects to California." Another offered: "Now, if our governor came in here with this project—or an oil and gas company—we'd welcome it. But the way you've presented it? The answer is no." One participant captured the central disconnect with a single, sincere question: "Why would we take perfectly good CO_2 and throw it away?" He'd spent his career at a geologic carbon dioxide production facility. To him, this wasn't waste. It was product.

This experience perfectly illustrated a major component of The Myth: "It's easy! Just explain the project and they'll want it." But it also hinted at The Myth's unraveling. One of the five conveners has since become a dear friend. She was moved by what she experienced that day—and quickly recognized how the assumptions she'd brought into that Wyoming room might actually *undermine* climate progress by deepening the very divisions we need to bridge.

But the other four left bewildered. They couldn't grasp why anyone would reject labor unions, community organizers, or environmental groups as partners in local development. The idea that carbon dioxide might be seen as valuable—or that climate action might feel like an external agenda imposed on a proud community—just didn't compute.

That's the power of The Myth.

As oil and gas leaders, we need to understand The Myth for three reasons: first, to see how it's unraveling—and where it still holds sway; second, to spot the opportunities it leaves in its wake; and third, to ensure it doesn't snap back even stronger, by offering a more compelling alternative.

We're not unpacking The Myth to mock it. We're here to understand its power—because only then can we lead effectively in a world where it's clearly not aligned with reality but there isn't a narrative (or set of pragmatic solutions) on offer to replace it.

Why Is The Myth So Potent?

Back in 2009, when I stepped into my role leading the Colorado Oil & Gas Association (COGA), natural gas was still considered a climate solution. Really. At the time, Governor Bill Ritter was championing his "new energy economy"—a vision that boldly included wind, solar, and yes, natural gas. Colorado's production was overwhelmingly natural gas, and the industry was starting to see itself as a partner in the clean-energy future.

And we weren't alone in that thinking. Across the country, the Sierra Club's "Beyond Coal" campaign was accepting more than $20 million from Chesapeake Energy—a natural gas company—to support its anti-coal work. Yes, you read that right: The environmental movement's flagship anti-coal campaign was, at one point, bankrolled by natural gas.

That alliance didn't last long.

But for a moment, industry leaders, environmentalists, and policymakers shared a pragmatic story—one that made space for diverse approaches to climate progress. The story didn't fall apart because it wasn't true; it fell apart because it couldn't survive the orthodoxy The Myth demanded. The very idea that oil and gas could play any role at all in solving the climate crisis quickly became heresy. The narrative shifted. The Myth hardened.

Over the next 15 years, the orthodoxy only grew stronger. Climate action became the dominant—and sometimes only—lens through which energy was viewed. What used to be energy policy and environmental policy quietly merged into one thing: climate policy. Priorities like water, wildlife, and land use were all folded under a single umbrella. Energy became a means to an end—and that end was climate.

It became an existential crisis, an apocalypse-in-waiting. And in that atmosphere of urgency, even basic questions became dangerous. If you brought up trade-offs or timeframes, you weren't being thoughtful—you were a denier.

This existential framing for climate ushered in a new kind of climate absolutism. You were either with the purveyors of The Myth or against them. There was no middle ground. No room for nuance. No safe space for the kind of open inquiry that good policy or leadership demands. And for oil and gas leaders, that meant we were expected to fall in line—or be cast as villains.

Eventually, even falling in line wasn't enough. Oil and gas companies were portrayed as insincere, untrustworthy, and ultimately incapable of meaningful climate action. In that story, there was no role for us except as the problem.

Understanding this backdrop isn't about blame; it's about clarity. The Myth is potent because it offers moral certainty in a chaotic world. But that certainty has too often come at the cost of real progress.

Many of you have read *All I Really Need to Know I Learned in Kindergarten*, by Robert Fulghum, so I hope the wisdom I opened this chapter with rings true: Imagination is stronger than knowledge. *Myth is powerful because it links our imagination to our identity.* When we believe a myth, what we *hope* will happen becomes more persuasive than science, logic, or even hard evidence to the contrary.

And that's exactly what we've been living through: The Myth of an Easy Energy Transition has held sway for so long—and over so many—because it's emotionally resonant, culturally reinforced, and built on a foundation of good intentions. The Myth has been attractive to so many people for so long for a few reasons:

It was simple and easy. The Myth's energy transition was one that would be straightforward, policy-driven, and affordable—all while minimizing personal sacrifice. In the conjured vision of The Myth, governments, industries, and the public would align around a rapid transformation to a cleaner energy future, powered by the right incentives and technological breakthroughs. If only we committed to this vision, The Myth suggested, we could move seamlessly from one era of energy to another—without disruption, trade-offs, costs, or hard decisions.

It had heroes and villains. If the transition was not happening quickly, there had to be someone to blame: Special interests! Greedy corporations! Corrupt politicians! This framing made it easier to believe that the very real emerging systemic challenges—engineering hurdles, capital requirements, and the need for reliable baseload energy—were simply roadblocks put up by bad actors rather than the inherent realities of a complex transition.

It had a great villain! The worst villain of all? Fossil fuel companies, of course! First, it was coal (which did play the villain rather well). Then, as coal use declined, the focus shifted to oil and gas (More on this in Chapter 6.) This villainization wasn't just rhetorical; it shaped policy, investment portfolios, and public perception. Oil and gas companies weren't seen as historical participants in the energy system—but as bad actors, peddlers of lies, and opponents of progress. The reality, of course, was far more nuanced. Oil and gas companies have been charged with powering the world and navigating the energy transition at the same time. But The Myth thrived on designating good guys and bad guys.

The result: Oil and gas leaders have been largely shut out of conversations about energy and climate solutions. The industry was treated as purveying the "fuels of the past," as an enemy of progress rather than an essential partner in building the energy future. Instead of being included in pragmatic discussions about emissions reductions, infrastructure, and innovation, oil and gas leaders found themselves on the outside looking in.

So Why Won't The Myth Just Disappear?

The perpetuators of The Myth weren't scheming masterminds. They were, for the most part, deeply concerned, well-intentioned people who genuinely believed a simple energy transition was possible. Policymakers, investors, activists, and academics saw a clear moral imperative—and assumed that if we just summoned enough political will, the technical and economic pathways would open before us.

Some were "for-profit activists," yes—building careers, businesses, or political influence on the idea that the energy transition was as easy as flipping a switch. But most were simply optimistic. They underestimated the scale. They overlooked the cost. They ignored the trade-offs.

And some of them are your family, your neighbors, even your kids' teachers. These are people whose identity is deeply wrapped up in the climate orthodoxy—and that makes it incredibly difficult to change their minds.

Matt Kolesar, chief environmental scientist at ExxonMobil, has become an important thought partner for me. We are both committed to engaging with climate-centric stakeholders, not only in service to our own understanding but also in service to progress. Here's what Matt told me about working with adherents of The Myth:

> *It comes down to trust and being open. I mean, these are tough conversations, right? There's a very strong emotional connection to people's beliefs that climate change is threatening their environment, which ExxonMobil and I personally very strongly agree with. So that automatically creates an emotional component to this. And so from that moment in a conversation, it's not just a science or engineering or math problem to negotiate. It's an emotional space to negotiate, and you have to be open and negotiate that up front.*

That emotional terrain—that attachment to identity, to righteousness, to being on the "right side of history"—is what gives The Myth its staying power.

In 2021, I was invited to co-host a conversation at a private weekend gathering—a small group of handpicked leaders interested in action on climate. This is the kind of gathering I love: a long evening, time for nuance, and the chance to genuinely listen, learn, and build unlikely alliances. I was there to represent the oil and gas industry. My co-host, a prominent voice in the climate movement, was there to represent the urgency of ending the use of fossil fuels. We were asked to give opening remarks, then guide the conversation.

I live for these moments, because there's often real possibility in the space between assumptions and understanding. But that possibility depends on shared rules of engagement—and that evening in 2021, my co-host hadn't come to listen. Instead, he had come to instruct.

We danced around our differences for the first hour—with me earnestly trying to find common ground. But every line of dialogue, every question, every statistic, even every human story I offered was batted away. Developing economies' need for affordable, reliable energy? Irrelevant. Hard-to-abate sectors? An excuse. The lives and livelihoods tied to energy access? A distraction. Eventually, we found the source of the rigidity: a carbon budget.

His worldview was built entirely on that frame. It was non-negotiable. A finite amount of carbon could be emitted to keep global warming under 1.5 degrees Celsius, and that budget was absolute. No matter the context. No matter the consequences. The conversation, I realized, wasn't a negotiation—it was a moral test.

As dessert was passed around, I shifted tactics. I asked him plainly, "What if you're wrong? What if countries, companies, and people burn more fossil fuels than the budget allows?" After all, if 1.5 degrees Celsius truly marks a hard boundary between survival and catastrophe, shouldn't we at least explore what happens if we cross it?

He didn't blink. "They won't. They can't," he said.

Over after-dinner drinks, I soberly reflected on my failure to make a point, any point. Reasonable guests offered a toast and said, "You tried." Hardly an endorsement of my efficacy. A bit later, when I sneaked out the back steps for a breath of fresh air, a server told me, "Oh man. That was brutal. But you came across as really nice."

At that point, I just bummed a smoke.

The next morning over coffee with my interlocutor, I tried again. "Setting an arbitrary limit on energy use doesn't stop people from exceeding it," I said. "Wouldn't it be wise to explore alternatives, mitigations, or the consequences of overshooting?"

He stared back at me, unmoved. "Only someone who works for the oil and gas industry would even ask such a thing."

That stopped me cold. I was frustrated—and I let it show. "I know you think we're evil," I snapped, "but we're actually trying to prevent exactly the same outcome that you are." *Shit. I probably should have had a cigarette before breakfast.*

He looked at me—genuinely puzzled. Not angry. Not indignant. Just … confused.

That moment has stayed with me. Because it revealed something deeper than disagreement. This was a man of intelligence and influence, a leader with followers, funders, and a platform. But he couldn't imagine a world where the carbon budget might be exceeded. Couldn't fathom engaging in that discussion. And now, couldn't conceive of oil and gas leaders as allies.

That's the power of The Myth. It doesn't just generate rigid beliefs—it goes on to build impenetrable walls around them.

As we'll explore further in Chapter 6, it's easy to conflate climate-concerned citizens with full-throated climate warriors. But that's a mistake—with real consequences. There's no upside to blaming well-meaning people—especially our stakeholders—for believing in and perpetuating The Myth. Most aren't ideologues like my dinner colleague; they're concerned, they're hopeful, and they've been told a compelling story with no room for dissent. Our job isn't to mock or scold them for holding on to it. It's to meet them with humility, invite them into a more complex conversation, and offer them a better story—one that's grounded in reality and rich with opportunity.

You'll be relieved to know I've gotten better at this since then.

So Why Does This Matter So Much?

This all matters tremendously, because The Myth has shaped everything—including our default responses to it. (I'll deal with that more in Chapter 3.)

As we'll explore in Chapter 4, oil and gas leaders need to understand The Myth to learn from its unraveling, to recognize its continued potency, and to spot the opportunities it leaves in its wake.

The Myth was comforting. It offered a story of swift, righteous progress that didn't require hard decisions, tough trade-offs, or inconvenient partnerships. To its believers, the energy transition could be willed into existence—by policies, protests, and pledges! But fictions like that have consequences. And The Myth's legacy lives on in ambitious climate commitments, enshrined in law and corporate strategy, that may take a decade or more to unwind and replace with commitments to pragmatic, workable solutions that deliver energy abundance while still addressing climate.

Leadership in The Moment demands openness to uncomfortable perspectives. Vicki Hollub, president and CEO of Occidental Petroleum, captured the situation perfectly when she shared this thought with me:

> *It's allowing yourself to be open to those who may tell you they want you to go away—having the ability to sit there and listen and to understand what they really mean, what they're trying to drive toward—that has made us a better partner.*

This sort of dialogue requires not just technical expertise but a willingness to engage without cynicism, listen carefully, and forge new paths forward.

That's why The Moment offers more than just a cautionary tale or a business opportunity for us. It's a call to understand the belief systems that shaped the world we're operating in—because those belief systems still influence your stakeholders, your employees, and your regulators.

The Myth failed the very people who believed in it. It excluded the expertise, capital, and problem-solving spirit that our industry brings to bear. It made oil and gas a villain to be vanquished rather than a partner to be engaged with. And in doing so, it slowed real energy and climate progress.

But here's the good news: As The Myth unravels, the space for new leadership expands. We don't need to argue with the past. We need to step into the future. Nile Garritson, the portfolio manager for sustainable investment and stewardship strategies at the California State Teachers' Retirement System, or CalSTRS, made a good point:

> *The scale of what some of these companies can bring to bear with respect to transition and decarbonization is unmatched anywhere in the world—in terms of the technological*

> *know-how, the size of the balance sheets and the cash flows supporting them, the scale of the organizations. There's a spirit, an attitude of problem-solving and optimism around being able to solve big challenges.*

Problem-solving optimism is what we're bringing to the table now.

The Myth is fading. The Moment is here. And with it, an invitation to reframe the conversation, reclaim our relevance, and rebuild the future with eyes wide open.

Let's make the most of it.

Next up: The Moment the unraveling invites.

THE MYTH: KEY TAKEAWAYS

- Two common energy-and-climate conversational traps:
  - "We are doomed, and we have to do everything within our power as quickly as possible to reduce climate emissions."
  - "Companies must lead the energy transition, and regulation should have forced them to act yesterday!"
- We need to understand The Myth for three reasons:
  - To see how it's unraveling—and understand where it still holds sway
  - To spot the opportunities created by its collapse
  - To ensure it doesn't snap back stronger because we failed to offer a compelling alternative
- What's The Myth's trajectory?
  - The Myth is unraveling—but it is not defeated.
  - We need to offer a viable alternative.

CHAPTER 2

The Moment

Our current form of politics here in the U.S. has become an impediment to building large-scale infrastructure.

—Alan Armstrong, Williams

I SPOKE WITH Alan Armstrong, then CEO of Williams Companies, in September 2024 about signs that The Myth was starting to unravel. At the time, I didn't expect the most telling sign to come from a Slack message at my own company the very next month!

As a Colorado resident, I've watched the state position itself as a climate leader—and, like all energy consumers, I'm concerned about grid reliability and rising costs. The Slack message linked to a news story about our governor raising those same concerns—and distancing Colorado from California, the undisputed leader of state climate ambition. Now *that* caught my attention.

Here's the story: In October 2024—just one month before Democrats were swept out of federal power—Colorado Governor Jared Polis (D) sent a letter to his own appointees, including leaders of the Public Utilities Commission and the Colorado Energy Office. In that letter, Polis urged the appointees to consider the consumer impacts of "beneficial electrification"—a core tenet of The Myth, often referred to as "Electrify everything." Most notably, he invoked California as an example he *didn't* want our state to emulate: "Other states that have not carefully managed the electrification transition are seeing large increases in utility rates. For example, in California average electric rates have roughly doubled over the past decade."[1]

Until then, Colorado's Democratic leaders had told only the story of an energy transition filled with benefits: lower costs, more jobs, and fewer emissions. Polis's letter was the first clear nod toward what I call the Three Ts of Energy Realism: trade-offs, timeframe, and truth. (More on those in Chapter 5.)

The Moment wasn't two years off—it had already arrived.

In a conversation with me, renowned political commentator and author of the book *One Billion Americans: The Case for Thinking Bigger* Matthew Yglesias captured the challenge succinctly, reflecting on Democrats' energy blind spots:

> *There's this tendency to cast the use of fossil fuels as if it's some kind of weird scheme that the producers foisted on us, but it's obvious ... When gasoline gets really expensive, people get annoyed. The reason people get annoyed is that it's a useful fuel for automobiles.*

Yglesias argues that the energy transition requires a more interdisciplinary approach—one that connects climate ambition with real-world economics and voter behavior. As he pointed out, not a single member of Congress cheered when gas prices spiked in 2021, even though it technically advanced climate goals. Why? Because people vote with their wallets, and they expect affordable energy—even if they care about the climate. As Ruy Teixeira, a nonresident senior fellow at the American Enterprise Institute, told me, "It's very clear, if you look at the data, that climate change is very much a tertiary issue for most voters, particularly working-class voters. They care most about cost and reliability."

That's the tension at the heart of The Moment: Voters want climate action, but not at any cost. The Myth told them the transition would be easy and free. Reality is proving otherwise.

What Is The Moment?

The Moment is result of a perfect storm of forces converging to unravel the Myth of an Easy Energy Transition. It's the time of growing realization—among political leaders, corporate decision-makers, and even climate-conscious citizens—that their ambitious climate goals are colliding with the hard realities of cost, infrastructure, and public tolerance. Well-meaning leaders who once championed net-zero targets without hesitation are now confronting a new, unavoidable landscape shaped by three realities:

- **Urgent pressure to meet real-world needs.** Energy must remain reliable and affordable—and those needs are now taking priority. In early 2025, New England faced high natural gas prices and limited supply during a cold snap, leading to increased reliance on oil and coal to meet electricity demand. This situation highlighted the region's vulnerability due to constrained natural gas infrastructure. State and local leaders, even those with strong climate commitments, are being forced to reconsider policies that risk blackouts or unaffordable heating bills.
- **The collapse of oversimplified narratives.** The notion that the energy transition required only political will has proven false. In California, the state government's aggressive push toward electrification has resulted in a tangled web of trade-offs—including mounting public frustration over sky-high electricity bills and increased fire risks tied to transmission infrastructure. Each energy source presents drawbacks, and leaders are now grappling with hard choices.
- **A constantly shifting landscape.** Technology, public opinion, and geopolitics are in flux. The United States saw this in 2022, when Russia's invasion of Ukraine reshaped global energy markets overnight, prompting President Joe Biden to release millions of barrels from the Strategic Petroleum Reserve—despite ongoing U.S. pledges to achieve decarbonization goals. Climate politics doesn't exist in a vacuum, and The Moment is defined by the constant push and pull between ambition and adaptation.

In short, The Moment reveals what The Myth concealed: Decarbonization is complex, political, and subject to disruption. And The Moment is now, when its complexity is now showing up in boardrooms, legislatures, and households across the country.

Crashing the Party, Loud as Hell

Just as we all got accustomed to the dominance of The Myth, The Moment crashed the party—spilling punch, flipping tables, and changing the playlist. It wasn't subtle, but it got our attention.

In the months leading up to the 2024 election, I had been tracking a set of 10 forces that were quietly undercutting the idea of a smooth, one-way trip to net zero. I thought this book would be an industry guide to a long, unpopular effort to convince climate-focused civic and business leaders to take these inconvenient truths (pun intended) seriously. Events such as rolling blackouts, price spikes in California, surging power demand from AI data centers, and Europe's unshakable reliance on natural gas weren't just little blips—they were flashing red lights. I figured I'd spend chapters helping you, the reader, build the case that these signals mattered.

Then November happened.

Once President Donald Trump was elected, The Moment didn't just crash the party—it kicked the door down. I found myself staring at the landscape, wondering if the decarbonization train had skipped hitting a curve and jumped the tracks entirely. Everything I'd been monitoring as early warning signs suddenly felt like yesterday's news. The front page was now filled with a very different story: blunt analysis of what the electorate actually cares about—and climate wasn't on the list. The Trump administration wasted no time rolling out an energy-dominance agenda: canceling funding for the National Climate Assessment and for federal decarbonization hubs for hydrogen, DAC, and CCS; issuing executive orders to boost fossil fuel development and challenge state climate regulations; laying off agency staff; and slapping tariffs on … well, everything.

But I kept digging into the bigger picture, and here's what I've come to understand: The forces that make up The Moment aren't inherently for or against decarbonization or an energy transition. Any one of them can act as a headwind or a tailwind, depending on the context. Take the 2024 election: Political forces didn't just shift; they reversed. Virtually overnight, the policy environment changed from one broadly supporting decarbonization to one actively questioning its value. But under a different administration, those same forces might have pushed the system in the opposite direction.

That's the thing about The Moment: It's real, but it's not fixed. *These forces are always in motion.* That means industry and civic leaders alike must learn to plan for volatility, not certainty. And plan to take advantage of The Moment while it's here—because it's a limited window of opportunity.

The best reaction to The Moment isn't to celebrate the reversal. It's to build strategy that seizes the opportunities The Moment presents and outlasts The Moment's catalyzing disruption. In the second half of this book, I'll detail how you should get started on that.

Sunday Morning: Smell the Coffee … or Grab a Mop?

The Moment may have shown up uninvited, but it also brought something to the party: opportunity. Not the fake, ribbon-cutting kind, but the real and messy kind that asks you, the leader, to show up differently. Amid the wreckage of The Myth, we have a rare opening to reframe the conversation, reclaim our relevance, and build the future.

Reframe the conversation. The old binaries—climate hero or villain, renewables or bust, leave-it-in-the-ground or apocalypse—no longer hold. The electorate isn't demanding purity; it's asking for answers. And while political winds will keep shifting, we can anchor the conversation in what matters: results. As Austin Knight, a vice president with Chevron New Energies, put it, "No one company, not a single country—nobody is going to solve this by themselves." That reality creates room for new ideas—many of which are already central to your business success. John Dabbar, a National Petroleum Council study chair, said it plainly: "You really want to have policies that are likely to be the same over the life of that investment, and having them swing wildly from one side to the other … It's really good to have them be consistent." Stability and pragmatism? Suddenly in style.

Reclaim our relevance. Our industry has been treated like a necessary evil—tolerated, not trusted. That's changing. But relevance won't come from defending the old playbook. Toby Rice, president and CEO of EQT, nailed it:

> *If as a society we don't simplify this very complex problem, we will continue to take splintered, non-cohesive actions … working on paper-straw solutions, things that make us feel good but have zero impact on actually moving the needle on climate.*

Our relevance comes not from being right but from helping our stakeholders focus on what matters most: reducing emissions, reliably and affordably.

Build the future. Want bipartisan urgency? Look at AI. Data centers, manufacturing, electrification—every next-gen ambition depends on abundant, affordable energy. As Williams's Armstrong warned, "We as a country are going to miss a huge opportunity that we may not recover from. … [I]f we can't provide low-cost, reliable power, we may very well lose this race." And the politics, he adds, "don't mesh very well with the long-term planning required to overcome this issue."

That's where we come in.

It is time to step into the leadership vacuum. But to do that, we have to resist the temptation to say "I told you so."

That's where we go next.

THE MOMENT: KEY TAKEAWAYS

- What created The Moment?
    - Pressing, real-world needs (energy availability, affordability, reliability)
    - Shattering simplicity (It's complicated!)
    - Constantly shifting landscape (technology, geopolitics, public opinion, politics)
- The Moment creates an opportunity.
    - Reframe the conversation—with complexity, not slogans.
    - Reclaim relevance—by delivering real solutions.
    - Shape the future—through consistent, constructive leadership.

CHAPTER 3

The Temptation

In war, you can be killed only once,
but in politics, many times.
—Winston Churchill

Pride goeth before destruction, and an
haughty spirit before a fall.
—Proverbs 16:18 (King James version)

IN MARCH 2022—less than a month after Russia invaded Ukraine—I was corralled in the corner of a conference hotel lobby during CERAWeek. A prominent CEO grabbed my sleeve and pulled me aside.

"Tisha! Everything has changed. What do you think this means for all this climate stuff? Net zero—is it over?"

He wasn't being flippant; he was genuinely asking. His company had taken bold, very public positions on emissions reductions—the kind that weren't exactly winning him popularity contests with his peers. Now he was wondering: Had the ground shifted? Was he about to be left exposed and alone?

Three years later, I still think about that conversation. Especially now, as leaders in tech, banking, and even energy trip over themselves to distance their brands from anything that might be labeled "woke." What I've come to realize is this: Feeling exposed is often just a matter of perspective.

Which leads me to The Temptation.

Whole Hog

For years, The Myth of an Easy Energy Transition shut our industry out. Oil and gas companies were painted as villains—excluded from the table, dismissed as irrelevant (or worse, obstructive). Our expertise, talent, capital, and practical decarbonization solutions were unwelcome. Promoting real, scalable energy progress in a climate-centric world? Nearly impossible.

Now, The Moment has arrived. And so has The Temptation—to default to a self-defeating response, in three steps:

- First, the impulse to declare total victory: "See? We told you so. This whole thing was never going to work."
- Then, the tidy narrative: The transition was always a fantasy, our critics were foolish, and we were right all along. This narrative feels good. Vindicating, even.
- But from there, it's just a short slide into absolutism: We stop listening. We stop bridging. We shut down dissent. And suddenly, we've become the very thing we spent a decade pushing back against—rigid, righteous, and stuck.

And that's how we squander The Moment.

Because right now, The Moment offers an extraordinary opportunity—if we don't blow it. And we *will* blow it if we use our new leverage to gloat instead of lead. For the first time in a long time, we have the space to reframe the energy conversation—around pragmatic, grounded solutions. Policymakers, advocates, even longtime critics are starting to see the need to bring our industry to the table. That means we have a chance to lead in building the energy future.

So what happens if we give in to The Temptation?

We don't have to guess—we've seen this movie before. Just look at the arc of the woke movement. What began as a well-meaning push for inclusion and justice created a cautionary tale about overreach, rigidity, and backlash. And now, some oil and gas supporters are headed down a similar path—with swagger.

They risk courting the return of The Myth. And that's one sequel no one wants to see.

Where Woke Went Wrong

Let's not confuse The Moment—this rare window of opportunity—with total victory or carte blanche to implement our wish list. Because as the rise and fall of wokeness shows, blowback for immoderation in a fundamentally centrist society is fast and fierce.

Where did woke go wrong? In four fatal and now familiar ways, each of which we're starting to mirror in oil and gas circles:

- **Too-simple narratives.** Under woke, complex challenges like climate became framed in absolute terms. The "obvious" answer—stop using fossil fuels today—was pitted against a cartoon villain: entrenched, greedy corporations.
- **Moralizing.** Those simplified stories recasting oil and gas professionals as bad actors meant that public support for the industry became radioactive—even among pragmatic Democrats.
- **Bridges too far.** What started as reasonable, incremental steps quickly became byzantine—complex, bureaucratic, and all-consuming. A good example: Sensible Securities and Exchange Commission climate disclosures ballooned into a wildly complex mess of Scope 3 emissions rules. A question we should ask ourselves often: When does the next logical step become a bridge too far?
- **Banishing dissent.** I'll be honest—I got swept up, too. There were times when I couldn't tell if I was being inclusive or just afraid to speak. Not on climate (that's my day job, and I've got plenty of courage there) but in other areas. That experience gave me a gut-level understanding of why the backlash has been so intense.

Now fast-forward to 2025. Look at the loudest voices in "anti-woke" politics and those waving the "Drill, baby, drill" banner. You'll see the same playbook: oversimplified stories, good guys versus bad guys, and an embrace of extremes.

And just like their predecessors', when the political winds shift—as they always do—those voices will be left exposed. Worse, they'll have proven climate hawks' most damaging critique: that oil and gas leaders were never serious about solutions for the energy future. That we were never sincere.

And that's why you're reading this book. Because you're a serious leader.

The Alternative: Lead with Grit and Conviction

So what's the alternative to falling for The Temptation? The leaders I work with who are thriving in this landscape channel the qualities that have always defined great oil and gas leadership—grit, resilience, conviction, and a deep commitment to problem-solving. We lean into strategies that align with the national mood of energy dominance while still delivering on community expectations and climate co-benefits. We pursue projects that make carbon intensity a competitive advantage. (I'll walk through these opportunities in more detail in Part II.)

Effective leaders don't flinch in the face of political wind shifts. They chart a course tied to long-term business goals, not fleeting headlines. If a leader feels exposed, it's usually because they were blown off course instead of anchoring to a durable strategy. (See Chapter 8 for tools to build that kind of strategy.)

It's tempting to romanticize leadership as pure courage, as standing up to the mob. But courage alone isn't enough to sustain leadership. Conviction, grounded in clear-eyed business planning, matters, too. ExxonMobil CEO Darren Woods explained on the Bloomberg Green *Zero: The Climate Race* podcast:

> *I'm, frankly, building a business and have a strategy that allows me to be successful in any scenario that you can forecast*

> *going forward. If you want to take the IEA [International Energy Agency] Net Zero, a very extreme decarbonization scenario—if you look at the businesses that we've established within the company, the core capabilities that we're building those businesses on—in that scenario, using the assumptions of the IEA, I grow earnings, and I grow cash flow. That's good for my shareholders. I'm responsible for doing that.*[2]

There will be times to pivot on strategy, yes. But what we saw company leaders do in early 2025 looked a lot like an oxygen-starved fish flopping about on land. That wasn't strategy—it was panic.

We seize The Moment by invoking what has always set this industry apart—our *conviction*—to solve the world's energy problems—the very DNA of our industry, sprung from our roots overcoming obstacles in the field. That's how oil and gas leaders operate. That's our heritage and our legacy: visionary leadership and can-do engineering.

Oil and gas field development is hard. Most North Americans are blissfully unaware of the nonstop, imaginative technical innovation required to produce the oil and gas they enjoy with such abundance. Every day, field personnel solve innumerable challenges while designing, siting, building, and operating oil and gas infrastructure. And that problem-solving defines who we are and what we do. As Albert Girgis, head of growth for New Ventures at EQT, described to me, oil and gas companies have a unique culture of grit and resilience that flows from our work in the field.

This isn't about leading the world back to a nostalgic past. It's about leading it forward—to a pragmatic, workable energy future. And yes, that future will include oil and gas. But it will also include CCS, hydrogen, community engagement, and real partnerships—yes, with those still clinging to The Myth. As ExxonMobil's Matt Kolesar told me:

> *Whatever you believed in, the energy transition was going to require significant policy change, change unprecedented in American history. The ultimate way to be successful is*

> *to engage—even with those who you think haven't been historical allies.*

In 2023, decarbonization looked inevitable. In 2025, it looks impossible. In 2028? It could go either way. And since we can't know which future is coming, a thoughtful executive prepares for both.

The Sh*t Hits the Flood

When I was running COGA, the Front Range of Colorado experienced a massive flood that isolated tens of thousands of people, required the helicopter evacuation of the most people (including my kids) since Hurricane Katrina, and buried thousands of wells under water. Activists hired helicopters, showed pictures of floating 55-gallon drums, and claimed that oil wells were compromised and leaking copiously. (Spoiler alert: They weren't. Not one.)

At COGA, we set up a communications center that engaged hourly with member companies to track assessment of wells and to interface with first responders, regulators, communities, and the media. And it won't surprise you that companies not only mobilized quickly, assertively, and comprehensively to assess and secure their operations, they also organized teams to deliver diapers and other necessities to isolated communities. They sent out staff and family volunteers with backhoes and shovels to clear neighborhood roads. For weeks, they led responses in the communities where they lived and operated.

That wasn't PR. That was conviction in action. That was resilience made real.

The Temptation's Last Gasp

"But they don't want *us* to solve it!"

You'll hear this from employees. From senior leaders. Even from the boardroom. It's the final breath of The Temptation—resentment dressed up as realism.

And it's not totally unfounded. Some stakeholders don't want oil and gas to be the heroes of the story. Or even *in* the story! They have spent years painting us as villains. And when you've been told—explicitly or implicitly—that your contributions aren't welcome, you're naturally going to bear a grudge. It's no wonder some in our industry now want to sit out The Moment.

Giving in to The Temptation would be bad business strategy, as I'll show in Chapter 4.

But also: Giving in to The Temptation is not who we are.

And: Giving in to The Temptation would supercharge the resurgence of The Myth—just when we have a golden opportunity to lead the world to something better.

We solve problems. Not because we're always invited, not because it's easy—because it's necessary. That's what leadership looks like in The Moment: showing up anyway. Refusing to play victim. Refusing to politicize progress.

Conviction, grit, and resilience aren't aspirational ideals for us—they're inherited traits. They're what built this industry. And now, it is on us to carry it forward.

In fact, if we don't evolve our mindset and our business strategies, we risk repeating exactly where woke went wrong: mistaking moral outrage for progress and turning complexity into purity tests.

And we know how that ends.

Lead Us (Not into Temptation)

As author James Pethokoukis says in his book *The Conservative Futurist*, "Our beliefs about the future matter, so let's act like it." Leadership today means sticking to your plan—even when critics are loud and allies are uncomfortable. Austin Knight of Chevron conveyed this spirit to me:

> *Stay curious. Have that growth mindset, because we don't know what we don't know. Stay excited. This is tough! Stay curious. Stay excited.*

We started this chapter in a Houston hotel lobby, with a CEO asking the question that many in our industry have quietly wondered: *Am I overexposed?*

That feeling—of uncertainty, of standing alone while the winds shift—comes to every leader. And when leaders feel exposed, they're most vulnerable to The Temptation: to declaring total victory, simplifying the story, and shutting out dissent. But we've seen how such reactionary thinking ends: Just look at the rise—and crash—of the woke movement. What began with good intentions lost its footing through moralizing, rigidity, and a refusal to entertain complexity. Some in oil and gas are now making the same mistakes: turning frustration into absolutism, grievance into a political platform.

We can fall for The Temptation—or we can rise to meet The Moment.

In the chapters that follow, we'll unpack the dynamics that created this opening, the strategic opportunity it presents for your business, and the practical steps for you to seize that opportunity.

THE TEMPTATION: KEY TAKEAWAYS

- Four tempting—but risky—responses to The Moment:
  - Declaring total victory—claiming "We won" and saying "I told you so."
  - Simplifying the story—creating neat, moralizing narratives that dismiss past efforts and depict opponents as naïve.
  - Silencing dissent—embracing rigidity, shutting down disagreement, and ignoring complexity.
  - Retreating—stepping back entirely, leaving the field open for The Myth to return unchallenged.
- The Moment creates an opportunity:
  - Reframe the conversation—with complexity, not slogans.
  - Reclaim relevance—by delivering real solutions.
  - Shape the future—through consistent, constructive leadership.

CHAPTER 4

The Opportunity

Someone's sitting in the shade today because someone planted a tree a long time ago.

—Warren Buffett

AT ITS CORE, The Moment is an opportunity—*but only for those willing to look past the obvious.* Yes, The Myth is unraveling. Yes, the world is once again hungry for cheap, reliable energy. Yes, oil and gas are back in the headlines for the "right" reasons.

But that's not where The Opportunity lies.

Plenty of industry leaders are busy making hay, doubling down on traditional plays to meet resurging demand. Sure, it feels good to be needed again. After you've spent years on defense, of course you'll be tempted to sit back and enjoy the return of pragmatism. But if you stop there, you'll miss what's actually happening.

The Opportunity is not about returning to business as usual. It's about recognizing that while The Myth may be crumbling, its believers don't have anywhere to go. Investors, policymakers, local communities—especially those who've staked their identity on climate action—are becoming disoriented. Some feel betrayed. The vision they held—of a swift, clean break from fossil fuels—is colliding with real-world constraints. (More on these stakeholders in Chapter 6.)

And here's the kicker: Those stakeholders still matter to you. They'll shape the next political pendulum swing. If you wait to engage with them until then, you'll be too late.

This is where The Opportunity lies for you—in three parts:

- **Reframe the climate-and-energy conversation** so it's not a zero-sum game.
- **Lead by example**—showing how energy abundance and decarbonization can coexist.
- **Build adaptive business lines** that can thrive under today's realities *and* adjust to tomorrow's.

And this isn't some future-state fantasy—it's already happening. A few examples:

- Carbon intensity is becoming a competitive advantage in LNG markets, where buyers increasingly prefer barrels with verifiable environmental attributes.
- Emissions reductions once framed purely as climate wins are now co-benefits tied to efficiency, permitting, and investor confidence.
- And community engagement—long seen as a "nice to have"—has become essential risk management, especially in jurisdictions where public trust determines project timelines.

None of this is new. But The Moment has raised the stakes. Leaders who once treated these sorts of efforts as side projects are now treating them as core strategy—not just to respond to today's pressures but to show climate stakeholders they're serious before those stakeholders are back in power and demanding receipts.

That's what smart oil and gas leaders have always done: Read the moment. Get ahead of it. Plant the tree. Avoiding The Temptation (Chapter 3) isn't about virtue signaling; it's about strategy. This is your chance to lead while others are still recalibrating. Because if you miss this chance, you may not get another.

Choose Your Own Adventure: You Determine the Future

Not everyone will embrace The Opportunity. Some companies will ride the fuel-of-the-past train as if it were the only one that mattered. And for a while, it might look as if they are winning.

But here's the thing about political winds: They always shift. Maybe in two years. Maybe in a decade. But eventually, climate concerns will once again take center stage—shaping capital markets, regulatory landscapes, and public sentiment in ways that will matter to your bottom line.

When that happens, will the progressive left—The Myth's loudest champions—be back in charge, returning with narratives sharpened by blame for lost time and setbacks? Will oil and gas be portrayed not as part of the solution but as the villain who stalled progress once again? In this future, you can bet that those who failed to demonstrate consistent commitment to decarbonization and innovation will find themselves marginalized. The companies that overplayed today's politics—who treated this moment like a hall pass on climate accountability—will find themselves on the outside looking in.

There's another potential future—one you create. In this choose-your-own-adventure, oil and gas leaders are proactive, reframing the climate-and-energy conversation as win-win, proving that energy growth and decarbonization can coexist, and building adaptive business lines that are strategic for both today and tomorrow. Here we lead our climate-concerned stakeholders into the energy future.

If your business depends on permits, local approvals, community partnerships, customer loyalty, or access to capital, you can't afford to ignore the future. Just a few years ago, shareholder resolutions were landing like thunderclaps. That storm may have passed for now, but the weather systems that created it haven't disappeared. They're dormant, not dead. And they'll be active again soon enough.

That's why leading in The Moment is so important, no matter what the conditions. It's about ensuring your relevance—today, and in a future where climate politics comes roaring back. That path leads to leadership—and you create it.

Let's Be Clear: This Isn't About Doing the Right Thing

"Climate action for companies is not about 'doing the right thing.'"

That's one of the most controversial things I say in public. (You should see the looks I get when I say it in an Enviro 101 class at the University of Colorado Boulder.) It drives the good lefties wild.

The same truth applies to The Moment. Acting on it isn't about virtue. It's about strategic clarity. It's about doing the right thing—for your business.

This isn't a game of timing the politics just right. It's about setting a course that advances your company's material interests over time—whether the climate conversation is center stage or in the wings. Ignoring what makes business sense just because it's out of fashion? That's foolish. Overcommitting to impossible expectations? Equally reckless.

Right now, The Temptation is strong. Your board, your investors, your employees are all wondering: *Why do the hard work of The Moment when decarbonization and sustainability seem to have fallen out of vogue?* Here's why: because real strategy means playing the long game.

There are three reasons why your company's climate and sustainability strategy should remain a priority, even when it comes to issues that were fashionable a year ago and wildly out of fashion today:

- **The political winds will shift.** Remember 2021, just after Biden took office? ESG was red-hot, net zero was a given, and every earnings call came with a climate explainer. It felt as if the world had changed overnight. That didn't last—and neither will today's narrative. Betting on permanence in politics is like betting on a newborn's sleep schedule.

- **The middle way is a hedge.** A steady, sensible course isn't indecision—it's risk management. It protects you from overexposure in any one direction and gives you room to maneuver when conditions change. Which they will.

- **Climate change poses material risks—and real opportunities.** The risks of a changing climate are well documented. Infrastructure, supply chains, insurance, even your future beach house for retirement—they're all in play. Oil and gas companies should be leading on emissions reductions within our purview—not out of guilt but because we can.

If you chase trends—or retreat from them—you miss the point. And worse, you miss the upside. There are novel opportunities ahead for companies bold enough to lead with both realism and ambition. But if you give in to The Temptation, you'll bounce from one moment to the next like a pinball—no strategy, no staying power.

You're Protecting Your Relationships and Reputation

Novel opportunities are emerging from two assets that cost little now but will be priceless later: relationships and reputation. You'll need both if you want to act on The Opportunity. That means tending to your climate-centric stakeholders today—even when it seems no one's watching. (Spoiler: They are.)

You don't have to look far back to see the payoff. During Trump's first term, companies that stood up for keeping methane regulations and made real investments in community-centered project siting didn't just earn praise—they earned position. When the Biden administration arrived, those same companies were the ones invited to shape new methane rules and lead hub consortia. Why? Because they had built credibility with environmental NGOs, academic partners, and local leaders—before they needed it.

Leaders who think in five-year increments tie sustainability to material business interests—not because it sounds good but because it makes good strategy. That doesn't mean placing bets on every shiny sustainability trend. It means being clear-eyed and consistent, focusing on what advances your core business while building the trust you'll need in the next cycle.

No matter which way the political winds blow, relationships and reputation hold their value. Think about what—and whom—you'll need 10 years from now. Then start showing up for them today.

These relationships and reputations will be critical not only in today's environment but in the inevitable return of intensified climate politics. Demonstrating consistent, pragmatic commitment now positions your organization to be part of the solution—not the problem—when The Myth rebounds. The goal isn't to win a war of narratives; it's to establish durable trust that survives political cycles.

Seizing The Opportunity: What Could Go Right

Let's flip the script. The Opportunity isn't just about avoiding missteps. It's about what you could build—right now—with the wind at your back. Scott Hallam, CEO of Boardwalk Pipelines, captured this spirit and vision when he spoke with me:

> *I'm confident that over the next five years, our industry will lead breakthroughs that reshape the energy conversation in ways we can't yet fully envision. We'll unlock new sources of value—opportunities invisible to us today—and these advancements will rally our stakeholders around a shared, compelling vision for the future.*

Here are three big advantages working in your favor:

A new energy–climate dynamic. In the eyes of climate-centric stakeholders, oil and gas leaders are suddenly the "winners" of today's political moment. That puts the burden—and the opportunity—of leadership on you. This is your chance to create a less polarized conversation, one that's honest about carbon emissions and bold about energy supply. You've already laid the groundwork through emissions reductions and decarbonization investments. Stay the course—and stay visible.

An opening for civic leadership. As The Myth of an Easy Energy Transition crumbles, climate stakeholders are looking for credible partners.

It likely won't be the original mythmakers they turn to—it could be you. You can step into the vacuum by showing that your commitment to both energy abundance and decarbonization isn't a phase—it's a strategic throughline. Show how the promises you made still hold, how your leadership extends into whatever comes next.

Resilient business lines. The oil and gas industry didn't endure for 150 years by being static. It thrived by evolving—technologically, commercially, and politically. That legacy continues. Innovation, shifting customer expectations, and political swings all require companies to adapt. The savviest leaders are already reducing carbon intensity, diversifying their portfolios, and investing in scalable decarbonization. These aren't side hustles. They're core to your future resilience. (More on this in Chapter 8).

EQT's Toby Rice has a uniquely pragmatic vantage point on The Opportunity. He understands how energy abundance can become a climate solution, not a contradiction:

> *Energy doesn't need to be the problem—it can be the solution. If we unleash the full potential of America's industry and resources, we can deliver affordable, reliable energy to the world and achieve massive emissions reductions at the same time. Our industry's innovation and optimism can tackle the world's defining challenges. That's the opportunity in front of us.*

That's the mindset The Moment calls for—and the one that will shape what comes next.

Acting on The Moment

The forces that unraveled The Myth have created something rare: a window, a chance to lead into an energy future that's more grounded, more durable—and still wide open. But beware: The Temptation is strong. You'll want to say "I told you so." You may even long for the simplicity of the more distant past—when oil and gas were valued without question.

But that past is gone. And this moment—The Moment? It won't last.

Understanding the forces gives you a roadmap—not just for what's happening now but for what comes next. This is your chance to lead with clarity, not complacency. In Part I, we've decoded The Moment. How do you turn a brief window into lasting advantage?

THE OPPORTUNITY: KEY TAKEAWAYS

- Where The Opportunity lies:
    - Reframe the conversation: Climate and energy are not zero-sum.
    - Lead by example: Energy growth and decarbonization can coexist.
    - Build adaptive business lines: Solutions can be responsive to both today's realities and tomorrow's expectations.
- The Opportunity in practice:
    - Carbon intensity as a competitive advantage
    - Emissions reductions delivering carbon co-benefits
    - Community engagement as strategic risk management
    - Long-term trust as a differentiator in future political cycles

PART II

Acting on The Moment

CHAPTER 5

The Choice

A true leader has the confidence to stand alone,
the courage to make tough decisions, and the
compassion to listen to the needs of others.

—Douglas MacArthur

IN PART I, I laid out how The Myth of an Easy Energy Transition came apart—and what's emerging in its place. We examined the gap between ambition and reality, between climate targets and the physical systems needed to reach them. And we named The Moment for what it is: a new phase, created not by idealism but by pressure—economic, political, and operational.

The unraveling of The Myth hasn't just exposed the complexity of the energy transition. It's revealed an opening for new leadership—leadership that's grounded in realism and equipped to navigate uncertainty.

That opening is The Opportunity. Not a return to business as usual—and not an occasion for shortcuts. It's the kind of opportunity that demands strategy, clarity, and the willingness to lead when others are still hoping someone else will go first.

Part II is about how to do just that—how to make this fleeting opening count. It's about what it takes to lead when the future isn't guaranteed and the playbook is still being written. The Moment will pass one way or another. The difference is whether your choice shapes what comes next.

Two Reads on The Moment

I'll be honest—at first I thought the urgency of The Moment would speak for itself. That a faltering grid, spiking costs, and climate goals slipping out of reach would lead to more pragmatic conversations about what achieving a functional energy future will actually require. I assumed civic and political leaders would start seeking out the people who truly know how to deliver.

I underestimated the staying power of The Myth.

Here's the tricky reality: Unraveling The Myth isn't the same as defeating it. It won't simply fade away quietly; it's deeply rooted and incredibly resilient. But the future isn't fixed—it's still wide open. And you have a critical role in shaping what comes next.

Like you, I hope we are moving past the oversimplified idea of a fast, painless transition. I want to see real engagement by the climate-concerned with the Three Ts of Energy Realism: timeframe, trade-offs, and truth. But we're not there yet.

So this chapter explores how we get there: what needs to shift for civic leaders to not just want those conversations but to want them with you. Because that's what's on the line: not just a better energy future but one in which you are invited to help lead it.

And let's not kid ourselves: That kind of leadership comes at a cost. You'll be criticized either way. The question is whether you'll be proud of what you choose—and whether it sets you and your company up for success now and later.

There are two ways to read The Moment. One says: The system is under pressure, so circle the wagons. Defend your legacy, attack renewables, and ride the wave of political backlash.

But understand this clearly: If you do that—if you sit this one out—The Myth wins by default. When the political tides inevitably turn and you've withheld your leadership, progressives—The Myth's loudest champions—will return with renewed urgency and ready to assign blame. They'll argue, "We lost precious time under Trump, and oil and gas stalled progress." Do nothing now and you'll face an even harsher landscape when The Myth rebounds.

The other way of reading The Moment says: The system is under pressure—finally, people are ready to listen. Step into the fray.

That's The Choice.

Choosing action over retreat isn't about winning today's news cycle or the current political tug of war. It's about cultivating the pragmatic middle ground, where real solutions are forged—resisting the temptation to either withdraw or lash out. It means stepping forward with partners, driven by pragmatism and purpose, demonstrating consistent commitment to innovation, decarbonization, and emissions reduction. That's how you shape the next chapter—not by dominating narratives but by earning lasting trust.

You can use The Moment to validate the past—or you can use it to shape what comes next. As Austin Knight of Chevron told me, "This is a moment like no other. We are crafting a new future."

The Problem-Solvers and Their Dilemma

The unraveling of The Myth has created real problems. Not theoretical ones—real, messy, high-stakes problems that someone now must solve. How you respond to those problems, and how you engage with the people responsible for solving them, says everything about your own strategy (and whether it will work).

The truth is, there's no single decision-maker who can fix what's breaking. Solving for a difficult energy transition is nobody's job—and everybody's job.

Take just one example: grid reliability. In any given state, it's not one person calling the shots. It's a shifting network of legislators, utility commissioners, regulators, utility executives, and staffers. Together, they will decide whether and how grid failures are addressed. But here's the catch: Every one of them answers to someone else. Legislators answer to voters. Commissioners answer to governors. Regulators answer to boards. And all of them are still, in some way, answering to The Myth.

That's the problem. And these are the problem-solvers.

They're the ones caught in the middle—tasked with keeping the lights on while navigating mandates that were built on oversimplified assumptions. Many know The Myth is unraveling. Some even understand what that means. But few have the political cover to say it out loud.

This is their dilemma: They need to deliver real results, but they can't afford to look as if they're walking away from climate ambition. They're expected to uphold climate-centric mandates—while quietly adjusting course to reflect physical and economic reality. And if they fail to show responsiveness to those mandates, they risk being replaced by someone who will.

So the burden falls on them—to challenge assumptions, communicate trade-offs, and propose timelines that actually add up. But they can't do it alone. To succeed, they need help from the very people The Myth has advised them to distrust: you.

That's your opening. That's when you make The Choice.

Because while they may not want to see oil and gas leaders as partners, they need what only you can offer: information rooted in operational truth, the courage to name hard realities, and the experience to co-design practical paths forward.

And here's the thing: If you want a seat at that table, you have to earn it. That means understanding the bind these leaders are in. It means skipping the "I told you so." And it means showing up not as a critic but as a collaborator. (Yes, even though The Temptation is standing in the corner with a smug look and a perfect one-liner. Resist.)

That's what The Choice really comes down to: Are you willing to lead, even when the people who need you most still see you as the villain?

The Three Ts of Energy Realism: Timeframe, Trade-offs, and Truth

So how do today's problem-solvers square the circle? How do they respond to stakeholders still clinging to easy answers—while also staring down real-world crises that make those answers impossible?

They start by reaching for the antidote to The Myth. And that's precisely where your expertise comes in.

Chad Zamarin, CEO of Williams, clearly captured why you want to make The Choice:

> *This is the time and this is the place where we can solve the world's greatest problems. We need energy—it's what makes the quality of life we enjoy possible—but we've got to decarbonize the energy ecosystem. There's no better way to do that than by leveraging our infrastructure and combining it with the technologies and energies of the future.*

Solving these complex problems won't happen overnight—there's no silver bullet. Instead, there are three deceptively simple concepts: timeframe, trade-offs, and truth. Together, they form a practical foundation for energy realism—and a common language for moving forward when the headlines start to blur and the pressure ratchets up.

Here's the critical step: communicating the antidote to The Myth. The problem-solvers must ultimately understand the Three Ts of Energy Realism and use them to frame meaningful conversations about today's energy realities and tomorrow's possibilities:

Timeframe. Remaking the energy system while also growing it isn't a five-year project; it's a generational one. That doesn't mean paralysis—it means embracing a planning horizon long enough to actually deliver results. Serious solutions require space for research, infrastructure, permitting, workforce development, and market uptake. Urgency isn't at odds with long timelines. Urgency just needs better planning.

Trade-offs. There is no free lunch in energy. Every source—fossil, renewable, or otherwise—comes with trade-offs: land use, emissions, intermittency, costs, impacts on communities, fuel sourcing, water needs, and on and on. Pretending some technologies are inherently good while others are bad isn't strategy—it's storytelling. If we want serious progress, we need to get technology-neutral and reality-based. And no, acknowledging trade-offs doesn't mean slowing down. It means stepping into the kind of leadership The Moment actually requires.

Truth. Eventually, someone has to say it out loud: Even with unlimited money and maximum mandates (hello, California), the transition would be expensive, complex, and resource-constrained. The physics are real. So are the permitting delays, the workforce gaps, and the technological unknowns. Leaders who want long-term credibility will have to tell the truth—even when it's uncomfortable. Because no matter how ambitious the goals, success depends on honest conversations with voters, customers, and policymakers about what's possible and what it costs.

The Three Ts aren't excuses. They're the foundation for serious, durable leadership—especially for those trying to keep the whole thing from falling apart.

There's just one problem.

You're Not the Ideal Messenger

Nobody wants to hear about the Three Ts from you.

And if they do, they're probably already aligned with you—which means they're not the ones you need to reach. To get into the real conversation, you'll need a more strategic approach (which we'll get into in the next chapter). But the first step is understanding why this conversation isn't already happening—why the Three Ts remain largely unspoken in most energy and climate circles.

- **There's no playbook.** Very few leaders have successfully navigated this terrain, and the ones who have done so have tended to do it quietly. There's not much precedent for being both pragmatic and praised.

- **There's no script.** Most problem-solvers don't have alternative solutions ready to go. And in today's polarized environment, even asking the right questions can get one accused of being a "climate denier."

- **There's no penalty—yet.** As long as the political and reputational costs of staying in The Myth are low, there's little incentive to step into hard truths.

So What's Your Lane?

Here's the paradox: Even though nobody wants to hear the Three Ts from you publicly, behind closed doors, many problem-solvers will take your call. And that's where the work begins. Alan Armstrong, now executive chairman of the Williams board of directors, explained it this way:

> *We don't have to make someone a winner or a loser out of this. There's a path forward that includes everyone, and we can actively support policymakers in navigating it successfully. For example, there are benefits to renewable energy that we can leverage alongside natural gas to build reliable, realistic solutions.*

Your most important task right now? Foster those quiet conversations. Be the reason these leaders start to build the courage ~~and cover~~ they'll need to lead with realism. Here's how:

- **Show up as their thought partner.** Understand the pressure they're under from climate-centric bosses and show that you respect what's at stake for them. Keep their climate goals at the center of the conversation—not as something to challenge but as something you can help them achieve more effectively.
- **Bring real ideas.** Show up prepared. Not with generalizations but with specific, actionable, climate-aligned solutions. When you understand their constraints, your experience becomes a source of strength—not suspicion.
- **Give them the credit.** These leaders won't be able to publicly elevate you as their guide. They'll need to find their own voice and their own allies. Your role is to support them.

Let's be honest: if the Three Ts start to show up in energy policy, it probably won't be because oil and gas leaders stood at the podium. It'll be because you were in the room when the problem-solvers needed a partner. Without your quiet influence, the conversation might not happen at all.

The Choice We All Face

So here it is. You can spend The Moment fighting for the status quo. Or you can be relentlessly future-focused—offering cleaner, more reliable, more affordable solutions and helping shape energy policy that reflects all three Ts.

Because The Choice isn't just about whether you lead—it's about *how* you lead. And whether you recognize just how strong The Myth still is for the people you need to reach.

If you've made the decision to lead, it's time to turn to the people who need you the most—even if they don't know it yet. Let's talk about your world of climate-centric stakeholders, The Partners.

THE CHOICE: KEY TAKEAWAYS

- The Choice before us:
    - Will you defend your legacy or proactively shape the future?
    - Step into the fray: Lead with pragmatism and innovation.
    - Resist complacency: Cultivate partnerships instead of retreating or lashing out.
- The Choice in practice:
    - Support problem-solvers caught between mandates and reality.
    - Engage stakeholders by articulating the Three Ts of Energy Realism: timeframe, trade-offs, and truth.
    - Quietly influence behind the scenes to build trust and practical pathways forward.
    - Engage proactively now to position your company strategically for the future.

CHAPTER 6

The Partners

If everybody is thinking alike, then somebody isn't thinking.

—George S. Patton

If there is any one secret of success, it lies in the ability to get the other person's point of view and see things from that person's angle as well as your own.

—Henry Ford

AS WE EXPLORED in Chapter 4, understanding The Moment isn't about "doing the right thing." It's about doing what's strategically right *for your business.* Your business exists within a vast world of climate-centric stakeholders. Some of these are the problem-solvers from Chapter 5; others include your investors, your customers, and your employees. Together they constitute The Partners.

These stakeholders are watching closely. When political power shifts, as it inevitably will, they will want accountability for lost progress, casting blame where it seems easiest: at your feet. But there's another path. You can proactively engage, becoming indispensable for the solutions your stakeholders seek. As Chad Zamarin of Williams offered: "We're not just going to talk the talk. We're going to walk the talk. We're going to show the solutions we can deliver as an industry to build the energy economy of the future."

Your Climate-Centric Stakeholders Are Not All the Same

At the heart of The Moment's opportunities are your climate-centric stakeholders. Some company leaders mistakenly think that the political winds have shifted permanently, so they're focusing on their traditional supporters. But as we discussed in Chapter 3, when did the wind ever shift permanently?

Last fall's elections dealt a blow to proponents of climate-centric energy policy. The trend was captured succinctly by an early-2025 poll covered by the *New York Times* under the title "Many Americans Say the Democratic Party Does Not Share Their Priorities."[3] In the poll, climate didn't even rank in the top five of respondents' concerns.

But that doesn't mean climate-centric voters disappeared. Far from it! In fact, in a December 2024 survey,[4] 73 percent of respondents said they think global warming is happening and 64 percent said they are at least "somewhat worried" about it. In your world of stakeholders, that's still a significant audience.

In just a few short years (or days, if you operate in politically blue jurisdictions), you will be making the case to climate-centric stakeholders about why they should engage with you or otherwise support your business endeavors. In short, they need to understand that they are The Partners.

I've been working with climate-centric stakeholders throughout my career, and here's how I categorize these audiences:

- **The climate warriors:** On one end of the spectrum, you have the 5 percent or so who take absolutist positions on energy and climate. Their positions sound like "Keep it in the ground," "There is no role for oil and gas companies in climate action," and "We need to get off oil and gas immediately." A sure tell of climate warriors: They are clearly more committed to the battle over climate than to making progress on emissions reductions.

- **The climate hawks:** This next 10 percent(-ish) is science- and data-driven. This cohort often starts with absolutist positions but is genuinely motivated by a desire for real emissions reductions. Climate hawks are characterized by sincere worry, intellectual curiosity, and receptivity to a wide range of solutions. (Note: Climate hawks do not all get along nor agree on acceptable solutions.)
- **The climate followers (aka the climate-curious):** This group represents the remaining roughly 85 percent of climate-conscious people. They often express climate-warrior-like positions, but these pronouncements are tied more to their peer group identity than to deeply held convictions. When it comes down to brass tacks (and copper pennies), they typically won't pay more for energy and its associated services. They will make largely symbolic (albeit sincere) gestures, such as giving up plastic straws or carpooling.

As an industry, we often mistake the large group of climate followers for climate warriors. Understanding the difference is crucial to our effective leadership in The Moment.

Most of Your Stakeholders Hold Their Climate Views Lightly

The most common error you are likely making: treating all climate-centric stakeholders as if they were climate warriors—dogmatic and unmovable.

I know from 20 years of engaging with these stakeholders that most of them hold their views lightly—much as you may hold your views lightly on health care policy or education reform. That leaves considerable room for meaningful engagement with roughly 95 percent of them.

Engaging with climate hawks is toughest for oil and gas leaders. Yet they are your most important future allies. Many problem-solvers fall into this group. If you're going to spend an outsized effort, spend it here. Understand what the world looks like to climate hawks, how The

Myth is driving their actions, and where you can contribute (sometimes quietly) to their solutions.

Amy Cradic, senior vice president and COO of New Jersey Resources, has worked in both politics and business. She reinforced this point to me clearly:

> *Whether you're on the government side of the policy discussion or the energy side, you must be willing to engage all stakeholders. You can't pick winners and losers from the start of the conversation. You'll lose something in translation if you don't understand everybody who's in this space.*

Working with climate followers is largely a matter of understanding their worldview, why they believe what they do, and whom they may trust to help develop their thinking. Because climate followers hold their views lightly, building trust by sharing authentic ambitions with them is an effective strategy.

Among your climate-hawk and climate-follower stakeholders, there has never been a better time than now to burnish your reputation. As the challenges of The Moment come to the fore, they are unsure of the trajectory to success. Remember: if they don't see alternative paths to progress, they risk sliding toward more rigid positions, like climate warriors'. (And The Myth will live on!)

Recall when the WTI price sank below $0 or when oil and gas company stocks were trading at a discount due to stranded-asset concerns? Relationships are forged during hard times. You remember who stood by you then, and loyalty endures.

How They See You

In my career, I have been repeatedly shocked by the vitriol aimed at me for working with the oil and gas industry. When I joined COGA as CEO in 2009, the fracking wars were just getting started. As a geologist, environmental scientist, and yogurt-making, cloth-diaper-wielding hippy mama, I genuinely believed I would be a compelling advocate for engaging with concerned environmental activists, community parents, regulators, and left-leaning legislators.

Boy, was I wrong.

Despite 15 years of environmental activism and science experience, I was immediately cast into the "tobacco" camp. It's a popular refrain, comparing oil and gas companies to tobacco companies: "You knew what you were doing was making us sick, but you kept doing it anyway." After years of hearing that charge, I grew desensitized—but today, I think it's important to unpack it together.

It's because, despite the fact that most of your stakeholders hold their climate views lightly, The Myth demands that people regard you as the villain in the climate morality play. You're going to have to work past that in many cases. Here's how.

A Tangent: It's Not Fair

This work gets easier once you accept the unfairness of it all. Schools teach your children that oil and gas is bad. Mainstream media publish one-sided stories about climate that make you the villain. My friends (and some of yours, too) assume your company wants to drill *whatever, wherever* with no regulatory constraints. Yup, that's all true. And not fair.

(Of course, given the current political environment, now isn't the time to cry a river to your climate-centric friends about how bad you've got it.)

Longtime readers know that I advocate consistently for taking the high road: Initiate conversations. Be patient. Maintain the effort to keep coming back to the topic.

I've now raised two young men, and I am proud that they grasp that life isn't fair. Rather than becoming bitter or reactive, they are inspired to be resilient and succeed anyway.

Life isn't fair, and the oil and gas industry has a lot to offer.

You as the Climate Villain

For a shortcut to how the world sees you: watch the 2005 movie *Thank You for Smoking*. When you sit down with a community member or regulator, there is a 50-50 chance that they see you as the oil and gas equivalent of a smooth-talking, story-twisting, wildly successful tobacco lobbyist.

Furthermore, anyone you identify as a credible spokesperson for your company—the local worker, the academic who led the study, the retired elected leader from the district—gets a very short window before they, too, are reclassified. They quickly, like you, are labeled as "a shill for the industry."

If you've had this experience, you know the feeling: a racing heartbeat, flushed skin, and a slight ringing in the ears. You might even feel it right now, reading this! We are hardwired to object to being typecast as villains. And though this reaction spurs our fight-or-flight response, reacting emotionally rarely changes minds.

Why care so much about thoughtful engagement? Because around 95 percent of climate-centric stakeholders are open to hearing you out. Savvy leaders won't make the mistake of treating all climate-centric stakeholders as climate warriors—dogmatic and unmovable. Trust me. I've been doing this work a long time.

So, what's a good oil and gas villain to do to win over The Partners?

- **Don't take it personally; take it as a challenge.** While it is frustrating to be typecast, this is simply a societal phenomenon. So rather than reacting, consider it a challenge! I often ask myself, *How long will it take for this person to be interested in what I have to say?* or *In what ways can I surprise this person and still be authentic?* There is little more satisfying than impressing someone who has underestimated you.

- **Learn their love language.** Read, watch, and listen to enough climate-centric material so you understand their shorthand. You don't have to invoke the "climate crisis"—I certainly don't—to be sensitive to their perspective and sense of urgency. And you can mindfully desensitize yourself to misinformation or attacks masquerading as news. Climate-centric stakeholders live and breathe this worldview—to engage effectively, you need to speak their language fluently.

- **Treat others as you ...** I am not asking you to lie or pretend people are right who aren't. I am asking you to indulge people in their misconceptions and biases. Do you wish others would give you the benefit of the doubt and not jump to conclusions? Let's do the same for our climate-centric stakeholders.

Here's why your next moves are so critical: Without proactive, pragmatic partnership from oil and gas, The Myth will surge back stronger and more hostile. Climate-centric stakeholders—the climate hawks and followers alike—need evidence of your genuine commitment today. Demonstrate that you are serious about partnership and innovation now—not later, after The Myth returns demanding stricter accountability. Showing up authentically and consistently right now is how you secure your seat at the table for tomorrow.

Working with The Partners

Investing now in the relationships and reputation you will need tomorrow is part of your durable, commonsense strategy. Below are some pro tips for engaging with climate-centric stakeholders:

Engage without blame. The Myth framed oil and gas as the villain, but its unraveling gives us a chance to reshape the conversation. By being relentlessly forward-focused, we can craft a compelling vision of our shared future, even though it won't be quick or simple. Cradic captured this in an example of how to do so with regulators:

> *We know that what we are operating are integrated systems—you have intermittent renewable energy sources alongside flexible fuels like natural gas. Start planning for the transition with reliability and customer impact first; if you start there, you'll end up in a pragmatic place where you can actually execute on decarbonization.*

Attack The Myth *gently*. Many well-intentioned people still hold on to all or part of The Myth. (Myths are powerful!) Rather than dismissing their perspectives outright, approach the conversation with patience and pragmatism. By meeting people where they are, we can replace The Myth with informed, solution-driven dialogue.

Lead with shared goals. The fall of The Myth doesn't mean abandoning climate goals; it means reshaping the energy dialogue. We can begin by acknowledging shared goals—lower emissions, energy security, and economic stability—while introducing the realities of scale, cost, and infrastructure needs. Don't wait for an invitation; step forward and lead the conversation proactively. Our North Star: shift our role from chief villain in The Myth to essential partner in building a sustainable energy future.

Articulate your vision. Current conversations around increasing energy demand have everyone (especially the problem-solvers) wondering who is going to step up to tackle this problem. We are! Step into the space with a coherent vision that acknowledges the opportunities and needs ahead and clearly articulates how your company and industry are already preparing. Cradic emphasized this approach: "This company has been leading from an opportunity perspective rather than from threats. We've leaned into innovation, demonstrating how our existing infrastructure can thrive in a new regulatory environment."

Now that you've got a clearer view of The Partners' perspectives and needs, let's examine the many situational factors at play in the next chapter: The Dynamics.

THE PARTNERS: KEY TAKEAWAYS

- Know your climate-centric stakeholders:
  - Differentiate among climate warriors (absolutist), climate hawks (data-driven, solution-oriented), and climate followers (climate-curious, identity-driven).
  - Recognize that most climate-centric stakeholders hold their climate views lightly, offering opportunity for meaningful engagement.
  - Accept and strategically address your default perception as a villain.
- Working with The Partners in practice:
  - Prioritize engagement with pragmatic climate hawks.
  - Learn their language and engage without assigning blame.
  - Lead with shared goals, patiently reframing The Myth through dialogue and solutions.
  - Consistently demonstrate genuine commitment to partnership and innovation.

CHAPTER 7

The Dynamics

I would not give a fig for the simplicity this side of complexity, but I would give my life for the simplicity on the other side of complexity.

—Oliver Wendell Holmes Jr.

MY UNDERSTANDING OF The Moment in 2025 is the end of a journey that started in 2010, when I made peace with working in oil and gas. I had just taken the helm at COGA, and I was out way over my idealism skis. I had left my comfortable environmental consulting job with a belief that I could do some good at COGA. It was a strange and short-lived window—one you may remember—when natural gas was being celebrated as a viable, even valuable, climate solution. Heck, even the Sierra Club was on board. Directional drilling and fracking had just transformed the game, and Colorado suddenly found itself sitting on a newly accessible, world-class oil reserve.

Suddenly, I was presiding over a Colorado oil boom. And I needed to understand what, exactly, I'd signed up for.

I had already examined natural gas and its role in the energy system—that's how I went from environmental activist to consultant to, as my elementary-school-aged kids called me at the time, "the president of oil and gas." So I decided to follow a similar course. One weekend, I got up before dawn, opened my laptop, and dug in. I made myself a deal: If, after reading and thinking and testing every assumption, I believed the world could move ahead without oil—or new drilling—I would find myself a new job. I've always kept myself in a position to walk away from any job if something doesn't sit right. This was a test of that principle.

You won't be shocked to learn that the opposite happened. My deep dive left me with even more conviction—not just about the role of oil and natural gas in powering modern life but about two new truths: (1) how important energy access is to the industrialization of developing economies and (2) the staggering scale of change required for meaningful climate action. These realizations did not leave me popular at parties in my liberal Boulder County enclave. But they did teach me the value of questioning and testing your ideas—again and again.

I've always been wired to chase the underlying truth. And I'm willing to upend my identity, my understanding, even my social circle to follow it. Which brings us here—to the work I do now. One of the most important parts is studying interplay between forces in the energy ecosystem—The Dynamics.

These experiences have made me unusually qualified to call balls and strikes in The Moment.

Ball: Many climate-centric thinkers are about to encounter so much anecdotal evidence and direct experience of how hard it is to decarbonize the energy system that even those with a very high bar for changing their minds will have to reckon with it. The Myth is powerful—but its power is about to be tested.

Strike: Being tested isn't defeat. Without a viable alternative to replace it, The Myth will resurge—perhaps even stronger—once the political winds shift.

Ball: *The Moment* was building for years. But then Trump was elected, and it felt as if it arrived all at once.

Strike: That doesn't mean today's dynamics are set in stone. Because dynamics are, you know, always *dynamic.*

What Is a Force?

Not long ago, climate-centric stakeholders widely believed the energy transition would be smooth sailing—driven by innovation, public will, and smart policy. That belief became so dominant that it reshaped the market: Energy stocks were devalued, and builders of critical infrastructure struggled to secure permits.

All that's changing.

Today, civic and political leaders are colliding head on with the realities of their net-zero ambitions. As we laid out in Chapter 2, those realities include three broad trends: urgent energy needs, the dispelling of the "easy transition" mirage, and a constantly shifting energy landscape.

So let's take a look at the forces driving the shift—what's bringing about the unraveling of The Myth and what you need to understand to navigate The Moment.

A ***force*** is an inherently neutral factor, something out in the world that, depending on the time and place, can either propel you forward or knock you back. Take decarbonization: When energy prices are low, consumers are more open to trying out new, lower-carbon options. But crank up those prices? Suddenly, even well-intentioned climate policies face fierce opposition. The same force—energy pricing—can act like a tailwind or a brick wall. It's not enough to know the forces shaping decarbonization. You have to know how they're behaving *right now,* in *your* situation.

So how do you navigate these forces? It starts with recognizing that no force acts in isolation—and none stays static. Each must be evaluated not just by its direction but by its strength, its speed, and the unique terrain it's encountering in your world. This isn't about predicting the future; it's about learning to read the weather so you can stay steady, no matter which way the wind blows. In Chapter 8, you'll see how companies are putting these insights into practice. Some are preparing for a more pragmatic future, while others are bracing for a return to climate purity. The wisest are building flexible, resilient strategies that take into account The Dynamics.

Ten Forces Shaping The Moment

As we sought to make sense of what was happening around us, my team at Adamantine zeroed in on 10 forces shaping The Moment—forces we believed were doing the most to speed up or slow down the energy transition. When we work with companies, we always start wide: We brainstorm a broad list of forces—things like "U.S. economy" and "geopolitics." That initial list might include dozens. But then we refine. We narrow it down to the few that really matter for the issue at hand. (I'll show you how to do that in Chapter 8.)

For this book, I focused on the forces most likely to move the needle right now—those with the power to accelerate or derail the transition in this unique time. These aren't abstractions. They're shaping your strategy, your reputation, and your ability to get things done. The future isn't locked in. How you and your company engage with these forces will help determine what comes next. Here are 10 of the forces most likely affecting your business right now and poised to shape its future.

1. **Grid reliability.** Rising power demand, limits on expansion and modernization, and shifts in generation sources are steadily undermining grid reliability. Organizations charged with monitoring the stability of the nation's grid, such as the North American Electric Reliability Corporation (NERC)[5] and Securing America's Future Energy (SAFE),[6] continue to raise the alarm through reports with such ominous titles as *Grid in Peril.*

2. **California energy prices.** California's energy and climate strategies are both celebrated by champions of climate policy and used in a cautionary tale for other states seeking to aggressively advance climate action. One of my Stanford colleagues likes to say, "California will break the energy system so the rest of us can learn what works and what doesn't." SAFE's *Grid in Peril* report also conveys the real challenges ahead for California's climate ambitions. As the

Center for Jobs and the Economy notes, in April 2024, California's average residential electricity price reached 30.16 cents per kilowatt-hour, nearly double the national average of 15.26 cents.[7] This significant disparity highlights the financial strain on consumers and the complexities involved in balancing ambitious climate goals with energy affordability.

3. **Data center demand for power.** Nothing has demonstrated the potency and speed of forces quite like the forecast of energy demand for data centers. Over the next decade, the rapidly increasing computational requirements and the deployment of AI are certain. What the data centers will require of the energy system is less so. Nevertheless, projections such as those in Goldman Sachs's recent report on AI, data centers, and the predicted U.S. demand surge[8] make this a central force to watch.

4. **Supply chain risk.** Supply chain risks are particularly acute across new-energy technologies predicted to be rapidly adopted. But supply chain risks, especially tariffs and trade wars, affect *all* supply chains, traditional and novel. Traditional fossil fuel sources have the advantage of established infrastructure and value chains as well as more diverse and flexible supply. This force will affect both the cost and speed of energy infrastructure expansions.[9]

5. **European demand.** European need for alternative sources of natural gas after Russia's invasion of Ukraine is a strong force demonstrating geopolitics' impact on The Moment. From emergency demand to keep the heat on to expected carbon-intensity standards, European demand for and requirements of energy imports will continue to serve as both an actual and a representative driver of global energy dynamics.[10]

6. **Changing demographics.** Once the baby boomers arrived on the scene in the 1940s, their generation was the driver of economic and political dynamics across the United States—until the millennials arrived en masse in the 1990s. The millennial generation will dominate the U.S. population in raw numbers through at least 2050, and their priorities and interests affect the rise and fall (and, surely, future rise) of interest in climate action. Global demographics will become increasingly relevant as populous emerging economies see the opportunity to move into the middle class (see last force!).[11]

7. **Policy and regulation.** Climate-centric scenarios still dominate the ways governments, investors, and analysts envision the future of the energy system. But on the ground, it's messier. In some regions—where The Myth is unraveling quickly—climate and energy policy are becoming increasingly uncoupled due to rising concerns over affordability, reliability, and geopolitical security. In others, The Myth still holds sway, and decarbonization remains the North Star. The result? A patchwork of policies. Some jurisdictions are accelerating mandates for electrification and emissions reductions, while others are rolling them back entirely. For companies, this means navigating a volatile policy landscape—where what's required (and rewarded) in one market could be penalized in another.

8. **The true cost of energy transition.** Decarbonization takes off when new fuels or technologies become more available, more convenient, or more cost-effective than the higher-carbon options they aim to replace. But over the past decade, progress has largely been policy-driven—and has therefore often come with a steep price tag for consumers. This force captures the uncomfortable truth: Well-meaning policies can push decarbonization forward while also driving up

energy costs. It encompasses such factors as carbon pricing, regulatory mandates, market imbalances, and shifting customer expectations.

9. **Technological advances.** Innovation can shift the playing field fast. New technologies may boost the competitiveness of oil and gas—by lowering costs, improving availability, or reducing carbon intensity. At the same time, breakthroughs in alternative fuels or energy systems could make them more attractive to consumers by making them cheaper, cleaner, or more accessible.

10. **Emerging economies' demand for power.** Low-income economies are also low-energy-consuming.[12] But as these economies grow, they'll need abundant, affordable, and reliable energy to fuel industrialization and lift populations into the middle class. Many currently rely on coal to meet power demand.[13] Whether the future favors lower-emissions sources will depend on a mix of economics, policy, geopolitics, and tech innovation.

How to Evaluate the Impact of the Forces

Understanding The Dynamics isn't a one-time exercise—you'll need to build the muscle and keep using it. The Moment offers a rare opportunity: Climate-centric leaders are starting to question long-held assumptions. They're opening to new inputs, new frameworks, and new partnerships. But if you want to meet them where they are—credibly and effectively—you need to keep tabs on what's shaping their thinking and why.

And remember: No political situation lasts forever. The winds are already shifting. Smart leaders aren't just reacting to today's conditions—they're scanning for what's next. Tracking the forces gives you early signals on where innovation, policy, and public opinion are heading. That's how you get ahead of the narrative instead of playing catch-up.

Yes, the forces differ across regions and sectors. But everywhere, they shape the strategic terrain. When you can read them, you can build a durable strategy fit that adapts in real time. That means spotting key signposts, making timely course corrections, and staying resilient when others get whiplash.

This future isn't set in stone. You're not just observing it—you're helping shape it.

If we don't shape what's next, The Myth snaps right back. Navigating The Moment demands more than reacting to change—it requires deliberate action, building new partnerships, and proving that the oil and gas industry is indispensable to real decarbonization.

So how do you stay abreast of the forces instead of getting blindsided by them? These four rules will keep you grounded and strategic.

Rule 1: Remember that forces are neutral—it's context that makes them vectors. A force is never inherently good or bad. Its impact depends on timing, geography, and what's going on around it. *Example:* Low energy prices can make it easier for consumers to adopt cleaner tech. But when prices spike, public support for climate policy often collapses.

Rule 2: Track interactions, not just existence. Spotting a single force isn't enough—you need to see how it's combining with others to shape outcomes. *Example:* Rising data center demand becomes a much bigger deal when it collides with grid constraints. That interaction makes it a political and reliability flashpoint.

Rule 3: Simplify the noise into signals. Tune out (but don't forget about) the parts that matter less to your business. *Example:* "Geopolitics" is broad and vague. But European gas demand after Russia's invasion of Ukraine? That's a clear signal—and it reshaped LNG markets overnight.

Rule 4: Plan for multiple futures. Forces evolve. The smartest strategies don't rely on a single forecast—they're built to flex. *Example:* California's aggressive climate mandates and Texas's market-first approach require two totally different long-term plays. If you're operating in both, your plan needs to flex.

We'll show you how to put these principles into action—and how to map your own force landscape—in Chapter 8.

Reading the Wind: Forces as Headwinds or Tailwinds

Depending on the time, place, and conditions on the ground, a force can either accelerate decarbonization or slow it to a crawl. In The Moment, the same force that once seemed like a clear win for climate progress can turn into a political liability or operational challenge. That's why it's helpful to think of each force at any moment in time as either a headwind or a tailwind. Remember (Rule 1): Forces are neutral—it's context that makes them strategic.

A headwind resists change. It pushes against decarbonization and props up the status quo. A tailwind supports momentum. It creates favorable conditions for accelerating carbon-intensity reductions.

Smart strategy means knowing which way the wind is blowing—and adjusting accordingly. That means watching for interactions between forces, not just their existence (Rule 2), and learning to spot specific signals amid all the noise (Rule 3). Let's look at two examples of how forces are shaping the dissolution of The Myth, the rise of The Moment, and what that means for business opportunity, risk, and building a durable strategy (as in Rule 4).

Headwind Example: Grid Reliability

In the spring of 2024, my local utility preemptively shut down power for the entire canyon where I live for an unspecified period of time, due to high winds, in order to prevent wildfires. At first, the absence of electricity was an inconvenience and a bit novel. It wasn't particularly cold, so the main considerations for me and my neighbors were things like going without wifi, keeping the refrigerator closed, and figuring out how to get water. (We are all on wells.) As the blackout stretched on, ultimately into a third day, the situation became increasingly dire. Food was spoiling. Elderly neighbors had no way to communicate with each other. (You could call only landline to landline.) Family members started calling 911 for welfare checks. Some residents needed power for their medical oxygen, which they hadn't had in days. People needed refrigerated medication and drinkable water.

By the time the power finally came back on, the mood had gone from patient to panicked.

Nothing shifts public sentiment faster than the power going out. Which is why reports warning of increasing grid instability—from NERC and others—should be taken seriously. The more Americans experience brownouts and blackouts, the less political oxygen there will be for any policy seen as contributing to the problem. (Rule 3 signal: Public tolerance for the energy transition is fragile.)

Grid reliability is a real and growing concern. NERC projects generation-capacity shortfalls across much of North America in the next five to 10 years.[14] And it's not just generation—transmission bottlenecks, aging infrastructure, wildfire risk, and explosive demand growth all contribute.[15] Some of these issues have nothing to do with the energy transition. But many do: Intermittent sources, mandated coal and gas retirements, and electrification policies are all stress-testing the system. (See Rule 2, about force interaction: Growing demand + retiring baseload = risk.)

As public anxiety over grid reliability grows, this force will act as a headwind to the energy transition—slowing its pace, reshaping policy, and shifting voter priorities toward reliability first. Your strategy needs to hold up even when political winds shift. (Remember Rule 4: Plan for multiple futures.)

Tailwind Example: Data Center Power Demand

Data centers are about to become one of the most powerful forces shaping the energy landscape. Demand is exploding, driven by AI, cloud computing, and big data. Whereas in past waves of tech infrastructure growth, companies merely scaled up, this time they have made serious, public climate commitments. (See Rule 3, on reading signals: Tech-sector growth meets climate action.)

That's what makes data center demand a tailwind. First, tech companies are actively pursuing low- and no-carbon power to meet their growth. Second, their appetite for energy is helping normalize and depoliticize new generation—whether it's clean, cleaner, or simply reliable. (This is explained by Rule 1: a neutral force, understood in context.)

Tech companies have made very aggressive climate pledges. For example, Amazon in 2019 announced a goal to match all electricity consumed through its global operations, including data centers, with 100 percent renewable energy by 2030.[16] Facebook (Meta) announced in 2021 that, although it had reached net-zero emissions by 2020, it would contract new solar photovoltaic and wind projects to remain at 100 percent renewable energy across its entire value chain.[17] In 2024, Microsoft updated its climate commitment to make data centers carbon-negative, water-positive, and zero-waste before 2030. Furthermore, it announced that, in 2025, it would procure 100 percent renewable energy for powering data centers on a global scale.[18]

In addition to pursuing renewable energy, many tech companies have been exploring applications for nuclear energy as a source for carbon-free power. In 2024, Google announced a partnership with Kairos Power, an energy-tech startup specializing in small modular nuclear reactors.[19] Amazon signed several deals supporting small modular reactors, as well, and announced plans to co-locate a data center next to an existing nuclear generating station in Pennsylvania.[20] Microsoft also entered into a power purchase agreement to prompt the restart of the Three Mile Island nuclear power plant.[21] These technology companies are clearly committed to securing the necessary resources to meet their growing appetite for energy.

Nuclear and renewable generating capacity can take a long time to develop. In the meantime, many companies have realized the need for natural gas to provide stable, reliable power to data centers. Project Stargate, a $500 billion proposed investment in U.S.-based AI infrastructure, supported by a consortium of tech companies including OpenAI, Microsoft, and Oracle, has sited its first data center location in Abilene, Texas, on a development that has its own natural gas generation capacity.[22] Chevron has also announced a partnership with Engine No. 1 and GE Vernova to develop natural gas generation facilities co-located with data centers.[23]

Data center demand is acting as a decarbonization tailwind—not by chasing climate policy but by chasing power. It's driving real investment in lower-carbon energy, building public support for new projects, and giving political cover for incremental progress—including the use of natural gas. (This is captured by Rule 2, on force interaction: Energy demand + innovation = tech acceleration.)

Expand Energy CEO Nick Dell'Osso articulated this industry imperative to me clearly:

> *The world is short on energy, and consumer demand continues to grow. Addressing this growing shortage is one of today's greatest global challenges. We are proud to answer the call for affordable, reliable, lower-carbon energy through our commitment to safely and efficiently delivering energy to markets in need.*

His perspective underscores the strategic importance of proactively aligning industry capabilities with emerging forces, to shape rather than simply respond to The Moment.

In other words: The tailwind might not be pure, but it's strong—and companies that harness it can ride it toward both business success and real sustainability. (And yes, your strategy should flex as such forces evolve, by planning for multiple futures, as in Rule 4.)

Making Sense of the Forces

The Moment is being shaped by The Dynamics—the play between forces influencing the pace, direction, and perception of the energy transition. Because forces are inherently neutral—their impact depending on context—they can act as with headwinds or tailwinds. In fact, the same force can act as a headwind or a tailwind, depending on your location, business line, or political environment. Your strategy must reflect the realities incumbent on each force.

Forces evolve. Smart leaders monitor the ones that matter most to their business, track how they interact, and build strategies that flex with the wind.

There's just one more thing to understand before you craft your own durable strategy: What is the full spectrum of opportunity available to your business in The Moment? That's what we explore next.

THE DYNAMICS: KEY TAKEAWAYS

- Understanding The Dynamics:
  - Forces shaping The Moment are neutral; their impact depends on timing, geography, and what else is going on.
  - Forces continuously interact, evolve, and shift, some even changing from headwinds (resisting change) to tailwinds (accelerating progress) or vice versa.
- Navigating The Dynamics in practice:
  - Track interactions among forces, not just individual forces.
  - Distill complex information into clear, actionable signals.
  - Build flexible strategies that anticipate multiple futures as forces evolve.
  - Proactively engage with forces to shape outcomes rather than merely reacting.

CHAPTER 8

The Approach

It ought to be remembered that there is nothing more difficult to take in hand, more perilous to conduct, or more uncertain in its success, than to take the lead in the introduction of a new order of things.

—Niccolò Machiavelli, *The Prince*

Plans are worthless, but planning is everything.

—Dwight D. Eisenhower

HOW CAN YOU FORMULATE durable plans amid intense pressure, volatility, and uncertainty? The story of PrairiePoint, a privately held oil and gas operator active in progressive states like Colorado and New Mexico, vividly illustrates this challenge.

So far, I have laid out why understanding the ongoing potency of The Myth allows you to see The Opportunity in its wake. In PrairiePoint's case, the company leaders did not fully understand that before them lay a critical juncture: The Choice. Following the 2024 elections, they initially felt compelled by shifting political currents to swiftly retreat from their ESG commitments.

Within months, this impulsive shift led to unintended consequences with their quickly reengaged climate-centric stakeholders. Key investors from Europe and Canada, who valued strong sustainability commitments, began questioning PrairiePoint's strategy. Meanwhile, state regulators still expected the best-in-class sustainability commitments from PrairiePoint that they had come to rely on. In a tense political environment, complicated by tariffs and trade wars, PrairiePoint was left scrambling to restore credibility and regain investor and regulator confidence.

PrairiePoint isn't a real company—I made it up. However, the situation it faced after a hasty strategic retrenchment is all too real. Companies invite risks by making reactionary decisions in response to political forces instead of evaluating them (and the opportunities they present) through a clear-eyed, thoughtful approach. Without such a strategic plan, the PrairiePoints of the world become trapped between urgent, conflicting expectations, highlighting a core lesson: Political tides inevitably shift, but durable strategies are built through anticipation and careful consideration.

To harness opportunities like The Moment, companies need a clear and focused plan of action, which they build through The Approach. An alternative futures analysis is the main strategy of The Approach, providing a framework to navigate the inevitable trade-offs of an uncertain future. This chapter sketches how you can use the alternative futures analysis and build such a framework: one that is grounded and robust, yet responsive.

Political winds shift frequently, and you can't lead by reacting to change. Instead, you lead through a carefully considered, durable strategy that anticipates rather than reacts to change. That's how you influence what comes next.

It's time to put your fingerprints on the future.

Everything Changes ... Including Today's Political Reality

From 2015 to 2025, the political winds shifted dramatically—not one, not two, but *three* times: from Obama to Trump 1 to Biden and back to Trump 2. These political swings reinforce a critical point: Tides will turn again, and when they do, the progressive left—the loudest champions of The Myth—will reassert their influence, likely returning with greater urgency and blame aimed squarely at oil and gas.

What does this tell us about our political fortunes in the next five years? The next 10? We know one thing for certain: Things will change. That leaves plenty of uncertainty: Where, how, and with what impact?

In the wake of the 2024 election, I spoke with company leaders who were faced with the decision whether to continue their investments in hydrogen hubs, transparent emissions data systems, and community benefits plans. They were under tremendous pressure—often contradictory pressures—to rationalize their investments. On the one hand, there was pressure to repudiate climate, DEI, and environmental justice endeavors. On the other, there was equally strong pressure to affirm their steadfast commitment to emissions reductions, community partnerships, and innovation.

It's no surprise, then, that one of the most pressing questions facing these leaders as they plan for the future is "What political environment will U.S. oil and gas companies face in 2030?"

Tackling this question (and other urgent questions posed by The Moment) requires a thoughtful, systematic process—one that helps you work through the multiple variables and outcomes in a way that channels anxiety productively. Arjun Murti, partner at Veriten and author of the *Super-Spiked* newsletter, explained it this way:

> *The fix-yourself trade is over. Balance sheets are better. Profitability is better. Now it's a question of how you fit into the world going forward. Do you actually have a positive view, or are you defensive? Too many companies are way too defensive right now.*

Let's look at the way I use alternative futures analysis with my clients.

The Alternative Futures Analysis

There are eight steps to conducting an alternative futures analysis. They can be done by a small team, either sequestered for a couple of days or in a series of one-hour meetings that build off each other. The point: to expand the team's aperture on both variables and outcomes and, ultimately, gather buy-in on recommended actions.

Step 1: Determine the question. A good question lays the foundation for the rest of the analysis. The question should be broad enough in scope that multiple scenarios are plausible, but specific enough to inform actionable strategies based on the outcomes. Right now, many companies are focusing on the question "What political environment will U.S. oil and gas companies face in 2030?" Other questions could explore public opinion, global security, or technological breakthroughs, for example. The analysis is conducted around just one question. Because the process is interactive, sometimes the question is tweaked mid-workshop. As Pierce Norton, CEO of ONEOK, told me, "One of the most important things in leadership is what questions we ask. It makes you slow down and gives you the chance to make better decisions."

Step 2: Choose your forces. Forces represent factors and dynamics that will affect the question's answer. Well-defined forces are neutral descriptions of these factors and dynamics—in categories such as economics, policy, society, technology, geography, and security. For the question "What political environment will U.S. oil and gas companies face in 2030?" forces could include geopolitics, the U.S. economy, the U.S. demographic composition, the political composition of the three branches of U.S. government, power grid capacity and reliability, energy technological breakthroughs, and cost of energy (fuels). To make the rest of the analysis process manageable, participants then distill the initial brainstormed list down to a list of no more than seven forces.

Step 3: Agree on the baseline. For each force on the list, the team articulates the baseline conditions: the status quo. The status quo is a description of how each force is acting right now. Coming to a common understanding of baseline conditions is an important step for a team charged with strategic decision-making. This step highlights differences of opinion on your team around current economic conditions, geopolitical influences, and the state of the regulatory environment.

Step 4: Map how the forces may act. For each force, the team builds a range of assumptions about how its behavior can change in the future.

A critical force to consider here is the resilience and potential resurgence of The Myth. For a force such as federal election outcomes, assumptions may include (a) all-Democratic rule, (b) all-Republican rule, (c) various configurations of divided rule, and (d) uncertain outcomes determined by the courts. Teams must acknowledge scenarios in which political changes allow The Myth to return with renewed vigor, directing blame toward the oil and gas sector, and complicating both operational and regulatory landscapes. Each assumption must be plausible within the timeframe of the question, should change only one variable, and should not overlap with other assumptions.

Step 5: Craft scenarios. With these building blocks in place, the team develops alternative future scenarios. In practice, these are often crafted independently (in a top-down fashion) by imagining plausible futures resulting from the range of forces considered. For teams with more time, they can be crafted in a bottom-up fashion derived from the mapping of forces. Alternative futures directly answer the question identified in Step 1. Examples addressing the question "What political environment will U.S. oil and gas companies face in 2030?" could include, for example, (a) the MAGA movement endures and dominates; (b) the political pendulum swings to the far left; (c) national politics undergoes a "great sorting," reflecting polarized politics by state; and, finally, (d) the country sees moderate, mixed results.

Step 6: Select signposts. Signposts are measurable, trackable, distinct indicators that the world is moving toward a specific scenario. For the current question, one signpost might be the price of gasoline, which can often foreshadow discontent with incumbent political parties. Another might be off-year election results in a state representing the company's political operating environment, foreshadowing election results in other, similar jurisdictions. Brainstorming signposts helps the team understand what to watch to determine which scenarios are becoming more likely over time and, thus, allows for the proactive adjustment of action plans.

Step 7: Mitigate risks and act on opportunities. Now that the team has envisioned various possible futures, it can identify the risks and opportunities each alternative future presents. For instance, an increasingly polarized political environment might bring significant regulatory swings and instability, whereas the continued rise of conservative ideology could bring even greater emphasis on energy security and domestic production. To keep the outputs manageable, the team agrees on suitable granularity, generally staying high-level. By assessing each future through the lens of underlying forces and corresponding assumptions, the team gains clarity on which paths are most fraught with risk and which hold the most promise.

Step 8: Identify options for action. With a clear understanding of the forces at play, the alternative futures on the horizon, and the risks and opportunities inherent in each, the analysis team is now well positioned to identify options for strategic action, such as maintaining or pulling back from investments, conducting outreach to new partners, or pursuing projects in a specific geography. The team can also prioritize approaches that could be successful in multiple scenarios—overlapping solutions that can help the company hedge in times of uncertainty. This step emphasizes creating options rather than deciding on a single path forward. Generating a range of options is crucial to successful decision-making: Garnering buy-in from the rest of leadership, investors, and stakeholders will require an astute assessment of options and the selection of some over others.

How Does This Play Out on the Ground?

Consider PrairiePoint, the exploration and production company I introduced earlier in this chapter. PrairiePoint represents a composite of several companies I've advised, combining salient qualities from each to illustrate how businesses can strategically navigate The Moment.

As a privately held oil and gas operator in progressive states like Colorado and New Mexico, PrairiePoint had been steadily reducing emissions, investing in low-methane completions, and pursuing net-zero Scope 1 and 2 emissions by 2035. These initiatives were central to the company's maintaining its social license to operate. The goal was clear: remain a competitive supplier in an increasingly decarbonizing market.

European capital and Canadian pension funds had enthusiastically backed the company's ESG-centric approach, and ESG had become its calling card. PrairiePoint even had a sustainability officer who had previously worked at an environmental NGO.

The 2024 presidential victory of Donald Trump dramatically shifted the political landscape, upending PrairiePoint's carefully constructed strategy. Suddenly, ESG was no longer just out of fashion—it was a political flashpoint. PrairiePoint's forward-thinking approach began to feel like a liability to the management team. Leaders grappled with the critical question of whether they should remove ESG from their public messaging and refocus strictly on traditional business fundamentals.

That question had no straightforward answer, because predicting the future is inherently uncertain.

The purpose of the alternative futures analysis, however, is to provide a structured framework for wrestling the possibilities into a set of actionable strategies. Rather than continuing their impulsive overcorrection, PrairiePoint's leaders convened a workshop to systematically evaluate their options.

Through this process, the leadership team at PrairiePoint identified three pathways to present to the board, each mapped to the opportunities and risks associated with possible future political environments. Because they couldn't control whether ESG would come back into fashion or not, the leaders created a strategy that would work in both scenarios. They reframed emissions efforts as operational excellence and cost control. They leaned into energy reliability and community engagement—language that resonates across the political spectrum. And they're still aiming for net zero—but now it's part of a broader narrative about being a resilient, trustworthy, long-term operator.

Delivering Gas to Communities amid Political Turbulence

BluePlains Energy, another composite case study, had long played the role of the hometown hero—delivering reliable, affordable natural gas to

Midwest communities, supporting local economies, and staying mostly out of the political fray. In the early 2010s, the company started exploring voluntary decarbonization: modest investments in renewable natural gas, some early pilots in hydrogen blending, and an internal task force to assess long-term emissions strategy. The idea was to quietly stay ahead of where policy might land.

Then came 2025, and not only did federal pressure to decarbonize evaporate but the whole notion of decarbonization became controversial. State regulators—eager to signal their support of the Trump administration—deprioritized climate and emissions endeavors. Meanwhile, in blue jurisdictions, "Electrify everything" advocates got louder and sharper—making "the future of natural gas" (aka "No future *for* natural gas!") a rallying cry to resist the federal policy environment.

BluePlains found itself squeezed, between climate hawks calling for gas bans in some jurisdictions, conservative stakeholders mocking its "out-of-touch" decarbonization investments, regulators demanding contradictory commitments, and customers just wanting their bills to stay low. Its leaders needed to decide whether to just hunker down, put their initiatives on hold, and wait for the storm clouds to pass.

Even before 2025, BluePlains was operating in split-screen environments with wildly different demands depending on the politics of the area. Yet stakeholders have no patience for a company that says and does one thing in one jurisdiction and something different in another. How could BluePlains succeed across its various operating environments at any time, let alone in the raised political tensions of 2025?

BluePlains got strategic about identifying these competing pressures in its operating areas. Leadership laid out two base cases describing the status quo operating conditions, one for blue-leaning and one for red-leaning jurisdictions.

Then BluePlains' leaders mapped out their alternative futures.

The alternative futures analysis process identified a range of futures that explored the extremes of today's operating environment. The company's leaders recognized that the key to identifying actions was to build a durable strategy that would acknowledge but not capitulate to their stakeholders' calls for gas bans. Simultaneously, every BluePlains consumer, even the most climate-conscious, prioritized energy reliability and affordability. So BluePlains reframed its initiatives as system modernization—producing best-in-class innovation, lower emissions, improved safety, and reduced costs. Leaders decided to frame decarbonization as innovation and customer choice, not ideology. They worked quietly with policymakers across the spectrum, making the case for a pragmatic "gas and" approach—gas and renewables, gas and low-carbon fuels, and gas and community partnership.

Gas for Export Markets

TrueVector Midstream, my final composite case study, made an early strategic bet: that international LNG buyers would start demanding lower-carbon molecules—and that U.S. exporters that could prove their emissions advantage would win in the global marketplace. TrueVector had ongoing methane monitoring, electrified compressors in key corridors, and piloted blockchain-based emissions tracking with upstream partners. Its thesis: Carbon intensity would become a price signal.

With the change in administration, TrueVector's leadership team faced critical questions: Domestically, would interest in emissions tracking wane, undermining their ongoing innovation and investment? Could TrueVector compete with companies rapidly removing their emissions-reductions commitments from their websites? Were global dynamics moving away from climate action and carbon intensity—and if so, was the move temporary or permanent? TrueVector had a significant first-mover advantage and its leaders wanted to maintain their investments, but board members questioned whether extending their sunk costs would ultimately prove unprofitable.

The alternative futures analysis workshop translated the seemingly binary choice before TrueVector into a wider range of options. Even as federal regulators deprioritized climate, many international buyers likely wouldn't. In a future scenario where European and Asian buyers required detailed GHG profile information, TrueVector's systems would prove advantageous. Under some economic scenarios, project finance teams also valued investments in carbon-intensity tracking systems. The company decided to focus on futures that balanced low regulatory pressure at home with rising carbon scrutiny abroad.

TrueVector leaned into two of the most likely futures. One scenario: International demand for low-carbon gas keeps rising. Another: Global pressure wanes under geopolitical shocks or energy crises. After weighing their options, company decision-makers built flexibility into their strategy. They created a differentiated "low-carbon-intensity corridor" for select export routes and secured a long-term, discounted third-party verification contract. They also ensured that their core assets remained competitive on cost alone, even without the carbon premium. They proactively engaged with U.S. policymakers to help shape—not just react to—emerging methane rules. Through the alternative futures analysis, TrueVector came to balance both possibilities: carbon intensity as a niche market signal today and also a passport to global relevance tomorrow.

Complexity Is a Feature, Not a Flaw

The Approach provides you with a blueprint for processing The Opportunity, which is presented by the unraveling of The Myth. The opening created isn't to return to the past or continue business as usual, but to craft a flexible, resilient business strategy. This requires understanding a range of possible futures, some of which may look a lot like a return to climate-centric priorities. Complexity is a feature, not a flaw in this strategic landscape!

Such complexity demands vigilance. If industry leaders remain passive or reactive and offer no alternative pathways forward for energy leadership, The Myth's resilience means it will return swiftly, reshaping public sentiment and policy priorities in ways that marginalize oil and gas companies even further. Only proactive partnership, clear vision, and pragmatic commitment can effectively change that trajectory.

Conducting an alternative futures analysis positions your company to actively seize leadership opportunities rather than merely react to external pressures. Every civic, political, or business leader awakening to the myriad challenges of climate action faces a crucial point—The Moment—in which to open a dialogue about how you, your company, and our industry lead with constructive solutions. As the pain points of The Myth's breakdown become more apparent and numerous, these conversations will create critical space to explore trade-offs and introduce pragmatic energy solutions and narratives.

In the next chapter, we look at how to step into the fray with your climate-centric stakeholders as proactive, credible partners. That's the only way to fully leverage and maximize the opportunity presented by The Moment. Ultimately, your goal should be clear: Shape the future—through consistent, constructive leadership.

THE APPROACH: KEY TAKEAWAYS

- Strategic durability comes from systematically evaluating multiple future scenarios rather than assuming one future will prevail.
- Engaging with uncertainty through structured planning allows leaders to proactively anticipate and navigate uncertainty instead of reacting impulsively.
- To conduct an alternative futures analysis:
    - Clearly define strategic questions about the future.
    - Identify key forces influencing outcomes.
    - Create scenarios reflecting a range of plausible futures.
    - Develop flexible, resilient strategies that can succeed across multiple scenarios.
    - Use measurable indicators, or signposts, to adjust plans as conditions evolve.

CHAPTER 9

The Case

You never really understand a person until you consider things from his point of view ... until you climb into his skin and walk around in it.

—Harper Lee, *To Kill a Mockingbird*

WHAT DOES IT MEAN when *the* governor for climate action suddenly cozies up to MAGA's biggest influencers?

Don't assume the answer is obvious.

There are still plenty of people living out The Myth. Even in states like my home state of Colorado, where early-2025 polling showed Democrats falling out of once-inevitable favor, climate concern is still the water within which we all swim. The Myth's influence is stubbornly persistent. As a result, your climate-centric stakeholders will have to be convinced that there's a viable alternative.

Which brings us to The Case.

As I wrote this book, I watched for signs that mainstream Democratic leaders might reposition themselves to the political and climate right of California Governor Gavin Newsom. I thought these shifts would signal that the Democrats' center of gravity was moving away from the progressives and toward a new, more powerful, pragmatic center—including the rise of practical environmentalism, which I'd been predicting since the November 2024 election.

Then in spring 2025, Newsom suddenly became the anti–Gavin Newsom, veering right on transgender issues and inviting the likes of Steve Bannon and Charlie Kirk as guests on his new podcast. This unexpected pivot signaled clearly that The Moment is even richer with opportunity than initially imagined. We'll return to this significant development later in the chapter.

Clearing the Path to Yes

Let's talk about the most underrated job of a leader: giving cover. If you want to see progress, you need to make it strategic for climate-centric leaders to say yes. That means having two things: first, a clear, credible company strategy that positions you as an indispensable leadership partner; and second, a coalition that's broad enough to weather scrutiny. If you want regulators, partners, or even your own employees to stick their heads up, you need to show them they won't get shot. No one moves alone—not in today's environment. People move when the risk is managed, the rewards are visible, and the path doesn't require self-immolation. That's cover. And providing it starts with your own house.

Remember, if industry leaders sit The Moment out, The Myth returns by default—more vigorous, more entrenched, and more punitive. Clear, credible leadership today doesn't just manage risk; it shapes the future narrative and prevents the conditions for The Myth's resurgence.

To effectively provide cover, oil and gas companies need to clearly articulate their role in meeting the world's energy needs. As Veriten's Arjun Murti emphasizes, "'Paris-aligned' is not a vision. That's defensive and defeatist. Articulate what you do, why it matters, and why people want your product." This proactive narrative is essential to clearly communicating your value to your partners.

The Mindset for Company Strategy

A strong strategy arises from being clear-eyed, humble, and willing to see the energy transition for what it really is: complicated, messy—and full of opportunity. You begin with ditching black-and-white thinking.

The old either/or frames—pro-economy or pro-environment, pro-oil or pro-renewables, climate champion or climate denier—aren't just tired. They're useless. If you want to lead, you have to break out of those binaries first. You can't help others evolve their thinking if you're stuck in your own orthodoxy. Simplicity might feel reassuring, but binary thinking isn't leadership—it's avoidance. Leaders who cling to the easy binaries miss critical strategic opportunities and fail to navigate the nuanced realities of The Moment.

This isn't about being vague or watering things down. It's about being strategic. The leaders who make the biggest impact are the ones who can hold complexity without getting lost in it. Complexity isn't a weakness—it's your advantage in The Moment. Just don't let your grasp of it turn into paralysis. People who know too much can get stuck. Your job is to stay agile.

As Chad Zamarin of Williams told me, "Urgency, optimism, and collaboration—these are essential. Creating connections across complex systems is the only way we will come up with big, sustainable solutions for the future."

To make The Case credibly, your company strategy has to be grounded in a dual commitment: both to decarbonize and to deliver affordable, reliable energy. That's not a talking point, and it's not negotiable. If you can't do both, your strategy won't hold up with the very stakeholders you seek to help.

Strategy Rooted in Humanity (aka Reality)

Too many strategies fall flat because they pretend the only thing that matters is data. "It's just math!" (or "physics!" or "science!"), people say, as if that settled the issue. But here's the truth: it's never just math. If it were, we'd have solved the energy transition a decade ago.

People make decisions based on fear, identity, loyalty, ambition—then justify those decisions with data. If your strategy doesn't account for those factors, it's incomplete.

This doesn't mean you throw out the facts. It means you lead with the truth of your stakeholders' humanity—their fear of being left behind, pride in being part of the solution, and belief that something better is possible. Strategies that focus only on data ignore the emotional and psychological drivers that underpin every meaningful decision. You can channel all of that. Emotion isn't a risk to manage—it's fuel. Use it.

And while you're at it, give people somewhere to go. If your strategy corners people, they'll fight. If it invites them in, they'll participate. Show them a role in the future. Make it real. Help them picture themselves in it. People move toward visions they can see themselves as part of, not toward abstractions. That's how change happens—not through pressure but through permission.

Start with what we agree on: The need for energy is growing. This reality doesn't have to be understood as a problem—it's a huge opportunity to shape the future energy system. More demand means more innovation, more investment, and more room to do things differently. Find that shared ground and build from there. It's not soft—it's strategic.

Durable Plans, Clear Vision

2025 reminded us that a strategy that changes with the political cycle isn't a strategy—it's a liability. If you want your team, your board, your partners, or your critics to take your leadership seriously, you need a vision that endures, one that shows where you're headed and your company's role in that future.

Everything you do—your decarbonization efforts, community investments, new tech plays—should further that vision. If it doesn't, cut it. Be ruthless about eliminating distractions. No one has time or capital for wobbly efforts that can't stand up to scrutiny.

And once you have that vision, tie it to objectives that endure. Not slogans, but real, material business priorities. When your strategy connects to what drives your business—access to capital, market competitiveness, regulatory stability—it becomes nonpartisan and nonnegotiable. That's how you build staying power.

Not every company needs a net-zero pledge or a billion-dollar carbon-centric tech fund. Every company does need a credible path forward—and specific solutions, which you can offer. That could mean

- Engaging stakeholders early and often
- Forging unexpected partnerships

- Leaning into R&D or new business models
- Getting your own house in order—on emissions, governance, land use, and water
- Managing reputation with real communications, not spin

When all these elements ladder up to a clear, compelling story of how your company will lead in the future of energy? That's when you make The Case. That's when people start saying yes. That's when coalitions form. That's when real progress happens.

To be a credible partner, ensure you have a public-facing strategy that articulates a clear vision, objectives, and elements that are consistent with your stakeholders' goals. This approach transforms strategic planning from a reactive exercise into proactive leadership, enabling you to build enduring partnerships and drive meaningful progress.

Building Coalitions

You can't do this alone—and neither can your stakeholders. The climate-centric leaders you're trying to support need more than encouragement. They need cover. And that doesn't come from a great strategy or bold internal action. It comes from a coalition: a diverse, durable network of partners who make bold leadership feel less like a risk and more like a movement.

Building coalitions is a concrete way to support the problem-solvers without placing yourself or your company at the front of the pack all the time. It allows you to engage with climate-centric stakeholders indirectly but powerfully, amplifying shared goals and diffusing potential backlash. Coalitions demonstrate broad consensus, legitimizing policy choices, reducing perceived risk, and increasing overall impact without positioning any single entity as the sole leader (or target).

The broader and more pragmatic your coalition, the more powerful it becomes. Think beyond the usual suspects. Climate advocates and clean-tech allies matter, but so do business organizations, large-scale energy consumers, unions, local chambers, and policy leaders across the political spectrum. When these voices come together, they give your stakeholders—whether they're CEOs, mayors, or agency heads—the ability to move. And sometimes, they even create the imperative to act!

Why? Because coalition members can invoke the Three Ts—truth, trade-offs, and timeframes—and still hold their ground. Having a coalition standing behind them is the difference between being exposed and being credible. Without this support, even leaders with good intentions can become isolated and ineffective.

But don't stop at advocacy. An excellent coalition isn't just for press conferences or op-eds. It must actively work to solve real problems—grid reliability, workforce challenges, permitting delays, and investment hurdles. When your coalition can actually deliver solutions—not just demands—you shift from being seen as an interest group to being indispensable. That's how lasting impact and meaningful progress happen.

For instance, consider a carbon dioxide pipeline and storage initiative that initially faced strong local resistance. With no natural constituency, the pipeline company built an unconventional coalition—uniting conservative farmers interested in pathways to new economic development, climate advocates driven by emissions goals, and influential community leaders focused on local resilience. This coalition didn't just offer political cover; it empowered climate-centric regulators to confidently champion the project's decarbonization impacts, while also providing conservative legislators with a compelling case for local economic growth. By bridging diverse interests, the coalition transformed a contentious proposal into a broadly supported, bipartisan success.

The most effective coalitions

- Include cross-sector and politically diverse members
- Focus on tangible outcomes, not abstract alignment
- Share risk, resources, and responsibility
- Understand that being pragmatic is not the opposite of being ambitious—it's the essential pathway to real progress

If your goal is to support bold leadership, build a coalition that transforms participants' perception of boldness from risky to inevitable.

The Scramble

Returning now to our narrative from the start of this chapter: The November 2024 election shook the Democratic establishment's sense of righteousness—and security. In the aftermath of a historic defeat, in which Democrats had seemingly misread every critical signal, climate became one of the earliest casualties. Amid this upheaval, the party quickly divided into factions: those doubling down on "the resistance," exemplified by the "Fighting Oligarchy" tour of Alexandria Ocasio-Cortez and Bernie Sanders, and those striving to create a broadly appealing, centrist alternative.

This alternative could offer a path forward distinct from both The Myth and MAGA—by appealing to a pragmatic majority. Consider Newsom's recent efforts to reposition himself. With his new podcast, Newsom was suddenly no longer the outspoken, climate-warrior governor of California, but was … what? Something different entirely. His initial conversations featured MAGA-aligned figures Charlie Kirk, Michael Savage, and Steve Bannon. The show was polished and professionally produced, representing more than just a podcast—it was a rebrand. Newsom tested this new space very publicly while also trying to hold on to his climate credentials. Indeed, in March 2025, he accepted a role as co-chair of America Is All In, described in its press release as a coalition of leaders working "to combat the climate crisis."[24]

It's a messaging mess—and sooner or later, your climate-centric stakeholders will be looking for clarity. That's something you can provide—because you're focused on The Opportunity, not trying to appeal to everyone at once.

And providing clarity is urgent—a credible, compelling alternative presented by pragmatic voices within oil and gas is necessary to keep The Myth from swiftly regaining dominance. Your leadership at this juncture will determine whether we revert to familiar adversarial narratives or forge something lasting and more constructive.

In many states where you operate, governors, legislators, and major corporations have already committed to climate goals. But these stakeholders are increasingly exposed at a moment when public patience for climate initiatives that jeopardize energy reliability and affordability is waning. (These decision-makers are precisely the problem-solvers highlighted in Chapter 5.)

Veritan's Murti underscored it for me:

> *These issues are far more complex than simply being in a climate crisis or it being overhyped. In the U.S. political context, both sides—left and right—stink. "Drill, baby, drill!" is as ineffective as "Keep it in the ground."*

Stakeholders who want to chart a more pragmatic path—but feel boxed in by past climate commitments or the persistent pull of The Myth—need certain things in order to move forward:

- **Permission to be pragmatic.** Problem-solvers who are climate-centric leaders need allies who value climate action but also acknowledge and normalize trade-offs. If they don't keep climate ambition front and center, the problem-solvers risk being labeled as sellouts by their constituents, so giving them permission to be pragmatic is important. They need to be armed with narratives and solutions of clear-eyed pragmatism lest they default to The Myth—which is still socially sanctioned. Your commitment to pragmatic leadership creates space for real solutions and prevents a return to polarized, counterproductive battles.

- **Narrative tools.** These climate-centric leaders also require trustworthy information that lays out clearly how climate action involves time and trade-offs. This information will be unique to their jurisdictions and convey the complexity of energy transition endeavors, with a focus on consumer costs, technical feasibility, and the critical imperative of maintaining energy system reliability.

- **Cover and more cover.** The broader the world of pragmatic stakeholders, the better. First, ensure you are a credible partner. Second, build coalitions of energy consumers, business organizations, and diverse political interests that enable stakeholders to invoke the Three Ts without being cast aside.

You can help your partners with all three of these needs. In doing so, you help them—and your employees—feel as if they belong to a shared endeavor. As ONEOK CEO Pierce Norton told me,

> *The first step is literally to say, "Okay, there's an issue." Recognize greenhouse gas emissions exist, and we have to do something to reduce them. Our employees feel they're part of something big—something bigger than themselves.*

Welcome to the Energy ~~Transition~~ Addition

Climate-centric leaders are facing an incoherence they can no longer ignore. They came of age politically and professionally believing the energy transition would be swift, simple, and morally unambiguous: powered entirely by renewables, driven by moral clarity, and enforced through political will. But now they're confronting an inconvenient reality.

So now what?

They're left holding two conflicting truths. First, the orthodoxies of the climate left still dominate the culture of their constituencies, shaping expectations and arguing for maximalist Myth. Second, governing—especially in The Moment—is forcing recognition of trade-offs, unpopular decisions, and a tolerance for complexity.

The orthodoxies of the climate left aren't just ideas—they're articles of faith. And they shape the way many climate-focused leaders, staff, and supporters see the world:

- **We're not moving fast enough.** The rallying cry of "More, faster, now!" has become a litmus test of climate commitment.
- **The energy transition is inevitable and unstoppable.** This idea coexists with the urgency of the first—creating a paradox: We must act immediately to accelerate something that will happen anyway.
- **Fear is the primary motivator.** The specter of climate apocalypse is used to silence debate, override opposition, vilify detractors, and therefore flatten nuance.
- **There are clear villains.** Fossil fuel companies, utilities, and political opponents are often cast as bad actors—making conversation, cooperation, or compromise with their expertise taboo.

These orthodoxies carry real-world consequences that extend far beyond messaging. They restrict the ways leaders operate, which ideas can be considered, and which partners are permitted. And they come at a cost:

- **They suppress innovation in the energy sector.** Energy stocks remain under pressure due to fears of obsolescence and divestment, even as energy demand grows.
- **They create stranded-asset risk—not just in infrastructure but in policy and permitting.** Agencies are approving projects that may not be viable. Legislators are advancing rules that ignore economic and technical realities.
- **They limit the leadership toolbox.** Pragmatic trade-offs—such as maintaining natural gas capacity or supporting nuclear—are off the table.

- **They foster black-and-white thinking.** Leaders are either climate champions or climate villains. This zero-sum worldview makes nuance politically perilous.

But the ground is shifting. New data, new constraints, and new demands from voters are revealing what many energy insiders have known for years: this isn't an energy *transition*. It's an energy *addition*.

Williams's Zamarin captured the pragmatic mindset perfectly:

> *We share a common dream with the people advocating for decarbonization. Let's use this moment to reaffirm we're not giving up on our shared goal. But let's acknowledge that the path we've been on hasn't been effective enough.*

The demand for reliable, affordable energy is growing, not shrinking. Renewables are an essential part of the future—but they're not replacing fossil fuels at the speed or scale imagined. Instead, we're layering new technologies onto an existing system. This is harder, slower, and more expensive than expected—and it comes with tough choices.

This complexity is not a failure. It's a feature. And it's the entry point for pragmatic leadership.

So how do you help climate-centric political or business leaders move from paralysis to leadership?

You give them permission to be pragmatic.

Narrative Tools

To be clear: Democratic leaders don't need to be "educated" by you on the primacy of oil and natural gas in our energy future! Instead, they need a way to explain to their supporters why climate action is still feasible, just more challenging than they anticipated. Here's how you can help them make The Case:

- **Speak their ambition.** Frame pragmatism not as a retreat from their climate aspirations but as the only viable way to achieve those aspirations in a changed political and economic environment.

- **Normalize trade-offs.** Talk about the energy system as a whole. Bring in examples from their states—stories of grid challenges, affordability pressures, or permitting delays—to illustrate why nuance is necessary.
- **Acknowledge the narrative pull.** Validate the ideals that brought them to this work, then gently guide them to newly emerging information that reshapes what leadership looks like now.
- **Name The Moment.** Many leaders feel this shift but don't know how to talk about it. Give them a name for what they're experiencing: "the pragmatist's pivot," "the post-transition reality," or "the next phase of climate leadership."
- **Celebrate courage.** Find and elevate those taking thoughtful, middle-of-the-road actions—especially in politically diverse states. Use their stories to create a new model of what a climate leader can be.

When climate-centric leaders are boxed in by outdated orthodoxies, they retreat—or, often, double down. But when they can see a path to leading with nuance, they move forward. And that's where The Moment will get its real, transformative power.

The Three T's—timeframe, trade-offs, and truth—are your superpowers. And they can be the superpowers of climate-centric leaders as well, if you give these leaders the tools to convey them. Here's how you can put these narrative tools to work with your climate-centric partners.

Timeframe. Climate leaders need political permission to acknowledge that the energy transition will take longer than their slogans (and prior commitments!) suggest. You can help them frame this not as backtracking but as realism—grounded in the infrastructure, workforce, and technology timelines we all share. Offer data, case studies, and operational milestones that show how steady progress still adds up to climate success, even over decades.

Trade-offs. Climate narratives leave no room for the messy middle, but that's where elected leaders live. You can provide the examples they need—of how gas reliability prevents blackouts, or how LNG exports support allies while building space for renewables—to explain why every step forward requires compromise. If you help frame these trade-offs in values-based language, leaders can communicate them to their own constituents.

Truth. Trust comes from saying the hard things clearly. Many climate leaders are already privately aware that some of their most ambitious targets are slipping out of reach and were never realistic. But they don't have the language to acknowledge this reality without sounding as if they'd given up on climate action. Share examples of progress that hold ambition and realism together. The goal isn't to lower expectations—it's to reset them, credibly.

The Moment calls for leadership grounded in realism, courage, and the willingness to embrace complexity. By giving climate-centric partners permission to speak pragmatically—and equipping them with narrative tools anchored in timeframe, trade-offs, and truth—you help create space for meaningful progress. The final chapter ahead will weave together all you've learned, outlining your path from navigating uncertainty to shaping a durable future.

THE CASE: KEY TAKEAWAYS

- Create conditions that enable climate-centric leaders to say yes.
- Provide your stakeholders with strategic cover by means of
  - A clear, credible company strategy that positions you as an indispensable partner
  - Broad coalitions that make bold decisions safer and more credible
- Apply these principles to your strategy:
  - Build strategies rooted in the real-world complexity of the energy transition instead of outdated binary frameworks.
  - Acknowledge the humanity and vulnerabilities of your stakeholders; they make decisions based on fear, identity, loyalty, and ambition.
  - Ground your company's approach in a vision that endures through political cycles.
  - Develop coalitions that span diverse political, business, and community interests, turning risk into collective action and lasting solutions.
- Your stakeholders need permission to be pragmatic!
- Give stakeholders the narrative tools and coalition cover to support their pragmatic leadership.

CHAPTER 10

The Close

The essence of work after providing for our simple survival is an intimacy between two seemingly opposing poles: an interior closeness to a foundational sense of self, and the felt longing for some recognized outer and as yet invisible horizon to which we dedicate our endeavors ... including especially, the physically felt, close-in invitations that first draw us, sometimes helplessly, to our calling.

—David Whyte, *Consolations*

I STARTED WRITING this book in ski season, and now I'm finishing it up during running season. This is a much more optimistic time when you live on the dark side of a canyon in the mountains! When I first sat down, I had no idea where this was all headed. The Myth of an Easy Energy Transition was obviously what I needed to write about, but what next? The incoming Trump administration seemed to throw all the cards of The Moment into the air!

That achy sense that things are changing fast remains. Yet the certainty that The Opportunity of a generation is upon us has replaced the foggy uncertainty of those initial days. Through the interviews, the thinking, the writing, and the revising, I became sure that the narrative is still being written.

And guess what? You're writing it now.

I think back to that dinner table in Colorado, where friendly conversation quickly turned into uncomfortable debate about climate and energy. That familiar discomfort signaled a realization: that the oversimplified story we've all been told was incomplete. That evening,

I sensed something critical was missing—nuance, pragmatism, and the courage to engage with the hard truths of what our energy future really requires.

Reflecting on that night now, I see how vital it remains to hold on to that early disorientation—to let it remind us of the complexities we face and the humility required to navigate them effectively.

Let's imagine what this conversation will look like four years from now. One scenario: absolutely zero progress, deeper partisan trenches, and you—once again—getting cast as the villain who single-handedly stalled climate progress during the Trump years. Frustratingly plausible, right?

It doesn't have to end that way. It's entirely your choice whether you seize The Opportunity that The Moment presents. Stepping into this opportunity means:

- **Reframing the climate-and-energy conversation** so it's not a zero-sum game.
- **Leading by example**—showing how energy abundance and decarbonization can coexist.
- **Building adaptive business lines** that can thrive under today's realities and tomorrow's expectations.

The rewards for such bold leadership will be immense—a new energy–climate dynamic, an opening for our civic leadership, and resilient businesses for the coming decades.

Lead and we determine history. And we consign The Myth to history, instead of watching it roar back to consume us all.

For each of us, we find our true work at the intersection of our innermost convictions and the compelling, yet uncertain horizon. The Moment calls you!

What comes next is up to you. Baton passed!

Endnotes

Chapter 2 notes:

1 Jared Polis to Rebecca White et al., October 8, 2024, https://completecolorado.com/wp-content/uploads/2024.10.08-re_-Beneficial-Electrification-Planning-for-Colorado-2.pdf.

Chapter 3 notes:

2 Akshat Rathi, host, *Zero: The Climate Race*, Episode 106, "What Is the ExxonMobil CEO Doing on a Climate Podcast?," Bloomberg, November 14, 2024, https://podcasts.apple.com/us/podcast/what-is-the-exxon-ceo-doing-on-a-climate-podcast/id1621556928?i=1000676941190.

Chapter 6 notes:

3 Jeremy Peters, Ruth Igielnek, and Lisa Lerer, "Many Americans Say the Democratic Party Does Not Share Their Priorities," *The New York Times*, February 2, 2025, https://www.nytimes.com/2025/02/02/us/democrats-ipsos-poll-abortion-lgbt.html?searchResultPosition=1.

4 Anthony Leiserowitz et al., *Climate Change in the American Mind: Beliefs & Attitudes, Fall 2024* (New Haven, CT: Yale Program on Climate Change Communication and the George Mason University Center for Climate Change Communication, 2025), https://climatecommunication.yale.edu/publications/climate-change-in-the-american-mind-beliefs-attitudes-fall-2024/toc/2/.

Chapter 7 notes:

5 North American Electric Reliability Corporation (NERC), *2024 Long-Term Reliability Assessment* (NERC, December 2024), https://www.nerc.com/pa/RAPA/ra/Reliability%20Assessments%20DL/NERC_Long%20Term%20Reliability%20Assessment_2024.pdf.

6 SAFE Grid Security Project, *Generational Growth: Grid in Peril: The Problem Statement* (Securing America's Energy Future, September 2023), https://safe2020.wpenginepowered.com/wp-content/uploads/2023/09/SAF-_GridSec_Sept23_v01.5_singles.pdf.

7 "California Energy Price Data for June 2024," Center for Jobs and the Economy, https://centerforjobs.org/wp-content/uploads/jun-2024-energy-report.pdf.

8 Carly Davenport et al., *AI, Data Centers and the Coming US Power Demand Surge* (Goldman Sachs, April 2024), https://www.goldmansachs.com/pdfs/insights/pages/generational-growth-ai-data-centers-and-the-coming-us-power-surge/report.pdf.

9 International Energy Agency (IEA), *Energy Technology Perspectives 2023* (IEA, 2023), https://iea.blob.core.windows.net/assets/a86b480e-2b03-4e25-bae1-da1395e0b620/EnergyTechnologyPerspectives2023.pdf.

10 IEA, *Global Gas Security Review 2024: Including the Gas Market Report Q4-2024* (IEA, 2024), https://iea.blob.core.windows.net/assets/fa115714-f9f8-4727-8520-5e8b5ca265ad/GlobalGasSecurityReview2024.pdf.

11 U.S. Bureau of the Census, *Demographic Turning Points for the United States: Population Projections for 2020 to 2060* (US Census Bureau, 2020), https://iea.blob.core.windows.net/assets/fa115714-f9f8-4727-8520-5e8b5ca265ad/GlobalGasSecurityReview2024.pdf.

12 Todd Moss and Hamna Tariq, *High-Income Low-Energy Countries Don't Exist* (Energy for Growth Hub, 2024), https://energyforgrowth.org/article/high-income-low-energy-countries-dont-exist-update-sep-2024/.

13 IEA, *Coal 2024: Analysis and Forecast to 2027* (IEA, 2024), https://www.iea.org/reports/coal-2024.

14 NERC, *2024 Long-Term Reliability Assessment.*

15 John D. Wilson and Zach Zimmerman, *The Era of Flat Power Demand Is Over* (Grid Strategies, December 2023), https://gridstrategiesllc.com/wp-content/uploads/2023/12/National-Load-Growth-Report-2023.pdf.

16 "Amazon meets 100% renewable energy goal 7 years early," July 10, 2024, https://www.aboutamazon.com/news/sustainability/amazon-renewable-energy-goal.

17 Urvi Parehk, "Achieving Our Goal: 100% Renewable Energy for Our Global Operations," April 14, 2021, https://tech.facebook.com/engineering/2021/4/renewable-energy/.

18 Noelle Walsh, "Microsoft's Datacenter Community Pledge: To Build and Operate Digital Infrastructure That Addresses Societal Challenges and Creates Benefits for Communities," June 2, 2024, https://blogs.microsoft.com/blog/2024/06/02/microsofts-datacenter-community-pledge-to-build-and-operate-digital-infrastructure-that-addresses-societal-challenges-and-creates-benefits-for-communities/.

19 Michael Terrell, "New Nuclear Clean Energy Agreement with Kairos Power," October 14, 2024, https://blog.google/outreach-initiatives/sustainability/google-kairos-power-nuclear-energy-agreement/.

20 "Amazon Signs Agreements for Innovative Nuclear Energy Projects to Address Growing Energy Demands," October 16, 2024, https://www.aboutamazon.com/news/sustainability/amazon-nuclear-small-modular-reactor-net-carbon-zero.

21 "Microsoft Deal Propels Three Mile Island Restart, with Key Permits Still Needed," Reuters, September 21, 2024, https://www.reuters.com/markets/deals/constellation-inks-power-supply-deal-with-microsoft-2024-09-20/.

22 "Announcing the Stargate Project," January 21, 2025, https://openai.com/index/announcing-the-stargate-project/; Ellen Thomas, "Stargate's First Data Center Is Underway in Texas. Public Filings Show How Much It Will Cost to Build," *Business Insider*, January 24, 2025, https://www.businessinsider.com/texas-stargate-data-center-build-cost-2025-1

23 "US Companies Make Bold Move to Power Nation's Data Centers," February 13, 2025, https://www.chevron.com/newsroom/2025/q1/us-companies-make-bold-move-to-power-nations-data-centers.

Chapter 9 Notes

24 America Is All In, "Governor Gavin Newsom Joins America Is All In as Newest Co-Chair," news release, March 21, 2025, https://www.americaisallin.com/governor-newsom-joins-america-is-all-in.

The Thanks

Gratitude is the understanding that many millions of things come together and live together and mesh together and breathe together in order for us to take even one more breath of air ...

—David Whyte, *Consolations*

THE JOURNEY OF writing *The Myth and The Moment* has been demanding, punctuated by critical moments of insight and doubt! I owe a particular debt of gratitude to those who affirmed my direction when I waivered and challenged my assumptions when I got cocky.

Sincere thanks belong to early interviewees Alan Armstrong and Toby Rice, whose conversations before I had begun in earnest convinced me that I was onto something important. Their willingness to engage candidly provided foundational confidence when I most needed it.

I am always grateful to Matt Kolesar, whose thoughtful challenges sharpen my ideas. To Amy Cradic, heartfelt thanks for regularly reminding me that this work was both relevant and necessary. Larry Larsen, you empower visionaries in a way that only a fellow visionary could.

This book benefited from the wise insights of Arnab Datta, Roger Pielke, Ruy Teixeira, and Matt Yglesias, whose expertise and candid (sometimes painful) observations offered nuance that enriched my understanding and therefore the narrative. To each of the experts whose insights appear throughout these pages—John Dabbar, Nile Garritson, Albert Girgis, Scott Hallam, Vicki Hollub, Austin Knight, Arjun Murti, Pierce Norton, Chad Zamarin—I express my profound appreciation.

My colleagues in the Real Decarbonization Consortium (some of whom are mentioned above) provided real-world stress testing of the ideas in this book. Thank you, Naomi Boness, Greg Blandford, Maria

Paula Camacho, Julie Dalzell, James Dunlap, Zach Keith, Joanne Mello, Carolina Santamaria Muñoz, Jaclyn Presnal, Kimberly Price, Jessica Roman, Michael Teague, and Maggi Young.

I couldn't have done this without my Adamantine colleagues. Becca Friedman, thank you for your tireless work on rigorous research and composite case studies! Morgan Gass, Maanav Jhatakia, Bevin Pan, and Jack Ridilla—thank you for your diligence, patience, and sharp analytical skills. Kayla Chieves brought remarkable energy to the podcast and content development process, offering essential support that made everything come together.

Bob Lalasz, you almost complimented me enough times that I found the courage to go on. You are relentless in your insights and high expectations of me. I'm pretty sure you owe me a table dance this time!

Caroline Schweiter, knowing that you genuinely enjoy reading my work has been one of the greatest compliments of this writing. Your enthusiasm inspired me more than you likely realize.

I could not have navigated the demands of this project without Anne Kurtis and Lindsey Slaughter, who regularly cover that little day job of running a company so I can pursue something knuckleheaded like writing another damn book. Your support made this endeavor feasible: I am grateful.

To Brian—I'm so lucky to have a life companion. Enough said, right? What the hell are we gonna do without a book project and any kids at home? (I've got a few ideas.)

Carter and Alec, you two are the purpose and the meaning. Study hard, get a job, don't move home. Love you!

About the Author

TISHA SCHULLER is an expert on the oil and gas industry's growing role in the global energy transition. As founder and CEO of Adamantine Energy, she advises oil and gas leaders on how to prepare for and take advantage of the transition's disruptions. Recognizing each company's unique position, she guides her firm's clients to develop tailored strategies based on their strengths—strategies that create significant transition-related opportunities. She specializes in honing policy positions, stakeholder engagement, outreach to unconventional allies, and community partnership. Her books—*Real Decarbonization: How Oil and Gas Companies Are Seizing the Low-Carbon Future* (2022), *The Gamechanger's Playbook: How Oil & Gas Leaders Thrive in an Era of Continuous Disruption* (2020), and *Accidentally Adamant* (2018) provide energy leaders with a detailed blueprint for effective leadership today.

Tisha also serves as the strategic advisor for Stanford University's Natural Gas Initiative and as an industry advisor for Avatar Innovations. Previously, she was president and CEO of the Colorado Oil & Gas Association and principal and vice president of Tetra Tech, a national environmental consulting and engineering firm. She has a bachelor of science degree from Stanford University.

Tisha serves on many academic and nonprofit boards, including those of the Breakthrough Institute, the Energy for Growth Hub, the Denver Museum of Nature & Science Institute for Science & Policy, the University of Wisconsin–Madison Nelson Institute for Environmental Studies, and the Payne Institute for Public Policy at the Colorado School of Mines. She is a member of the National Petroleum Council, an advisory board to the U.S. secretary of energy under the Obama, Biden, and Trump administrations.